Palgrave Studies in Media and Environmental Communication

Series Editors
Anders Hansen
School of Media, Communication and Sociology
University of Leicester
Leicester, UK

Steve Depoe
McMicken College of Arts and Sciences
University of Cincinnati
Cincinnati, USA

Drawing on both leading and emerging scholars of environmental communication, the Palgrave Studies in Media and Environmental Communication Series features books on the key roles of media and communication processes in relation to a broad range of global as well as national/local environmental issues, crises and disasters. Characteristic of the cross-disciplinary nature of environmental communication, the books showcase a broad variety of theories, methods and perspectives for the study of media and communication processes regarding the environment. Common to these is the endeavour to describe, analyse, understand and explain the centrality of media and communication processes to public and political action on the environment.

Madison P. Jones

Inventing Ecosystems

The Rhetoric of Science in an Ecological Age

Madison P. Jones
University of Rhode Island
Kingston, RI, USA

ISSN 2634-6451   ISSN 2634-646X (electronic)
Palgrave Studies in Media and Environmental Communication
ISBN 978-3-031-98792-2   ISBN 978-3-031-98793-9 (eBook)
https://doi.org/10.1007/978-3-031-98793-9

This work was supported by URI Office of Extension and Agricultural Programs and Center for Humanities, University of Rhode Island.

© The Editor(s) (if applicable) and The Author(s), under exclusive license to Springer Nature Switzerland AG 2026. This book is an open access publication.

Open Access This book is licensed under the terms of the Creative Commons Attribution-NonCommercial-NoDerivatives 4.0 International License (http://creativecommons.org/licenses/by-nc-nd/4.0/), which permits any noncommercial use, sharing, distribution and reproduction in any medium or format, as long as you give appropriate credit to the original author(s) and the source, provide a link to the Creative Commons license and indicate if you modified the licensed material. You do not have permission under this license to share adapted material derived from this book or parts of it.
The images or other third party material in this book are included in the book's Creative Commons license, unless indicated otherwise in a credit line to the material. If material is not included in the book's Creative Commons license and your intended use is not permitted by statutory regulation or exceeds the permitted use, you will need to obtain permission directly from the copyright holder.
This work is subject to copyright. All commercial rights are reserved by the author(s), whether the whole or part of the material is concerned, specifically the rights of translation, reprinting, reuse of illustrations, recitation, broadcasting, reproduction on microfilms or in any other physical way, and transmission or information storage and retrieval, electronic adaptation, computer software, or by similar or dissimilar methodology now known or hereafter developed. Regarding these commercial rights a non-exclusive license has been granted to the publisher.
The use of general descriptive names, registered names, trademarks, service marks, etc. in this publication does not imply, even in the absence of a specific statement, that such names are exempt from the relevant protective laws and regulations and therefore free for general use.
The publisher, the authors and the editors are safe to assume that the advice and information in this book are believed to be true and accurate at the date of publication. Neither the publisher nor the authors or the editors give a warranty, expressed or implied, with respect to the material contained herein or for any errors or omissions that may have been made. The publisher remains neutral with regard to jurisdictional claims in published maps and institutional affiliations.

This Palgrave Macmillan imprint is published by the registered company Springer Nature Switzerland AG.
The registered company address is: Gewerbestrasse 11, 6330 Cham, Switzerland

If disposing of this product, please recycle the paper.

One thought ever at the fore—
That in the Divine Ship, the World, breasting Time and Space,
All Peoples of the globe together sail, sail the same voyage,
Are bound to the same destination.
—Walt Whitman, Leaves of Grass

I read the Dr. Bronner's bottle
Earth's a spaceship and rinse
—Jesse Welles, "Life is Good"

Acknowledgments

As luck would have it, no one writes alone. Like all books, this project was conceived, developed, researched, sustained, and written within the context of several different intellectual ecologies and over many years. This book has benefitted from countless discussions and collaborations, as well as encouragement and feedback, from numerous friends, students, community partners, and colleagues across rhetoric, writing, and communication studies, the environmental humanities, and in the environmental and life sciences.

I was inspired to undertake a project on Howard T. Odum and ecosystems ecology as a graduate student studying rhetoric at the University of Florida (UF), working with Sidney I. Dobrin, mentor extraordinaire. There, I first imagined how H. T. Odum's work might inform the theories, methods, and practices of rhetoric, writing, and communication studies. I want to thank my graduate committee members, Sid, Raúl Sánchez, Anastasia Ulanowicz, and Robert Walker, who each offered keen advice in the early stages of inquiry that led me to this project. Many of the lines of thought that brought about this book began with coursework in their classes and through numerous conversations about the dissertation. They encouraged me to pursue research with Odum's work and each brought connections between ecology, information theory, decolonial work, actor-network theory, and place-based research to the forefront of my thinking. Heartfelt thanks to Cynthia Barnett who taught me so much about Florida's important role in the history of environmentalism and showed me how great teachers can take learning beyond the traditional borders of the classroom. I owe additional debts to the generosity of the faculty and graduate students in the English department at UF, and to my fellow members of the TRACE Innovation Initiative, especially Jacob Greene, Shannon Butts, Chloe Anna Milligan, and Aaron Beveridge, as well as the department staff, Melissa Davis in particular, who always helped me work through the complex bureaucratic systems that so often constrain research.

After graduating, I moved to the University of Rhode Island (URI), where new places, environments, and people continue to inspire me. My dual appointment in the departments of Professional & Public Writing (WRT) and Natural Resources Science (NRS) has provided me with the incredible opportunity to collaborate with tremendous and supportive colleagues across what C.P. Snow (derisively) called the "two cultures" of academia. You have all shown me the value and necessity of interdisciplinary inquiry and the value of bridging divides. Thank you to Genoa Shepley, our fearless WRT chair, who advocated for research time that supported writing this book and helped me articulate its value to funders; to Jeremiah Dyehouse, who gave incisive and meaningful feedback across many stages of developing this project; to Leah Heilig, Stephanie West-Puckett, and Lehua Ledbetter whose collaboration and support helped sustain the work of this book; and to Donna Hayden, our department admin, who helped me navigate the paperwork that made research travel possible. To my colleagues and collaborators in NRS—especially Scott McWilliams, faculty mentor *par excellence*—and Art Gold, Nancy Karraker, Sunshine Menezes, Peter Paton, Alissa H. Cox, Tolani Olagundoye, Michelle Peach, Jason Parent, Brett Still, and Graham Forrester—thank you for always making me feel welcome in an interdisciplinary space. Thank you to my collaborators across the university, especially to Ingrid Lofgren for hosting biweekly writing times that made space for steady progress on the manuscript, and to Amelia Moore, Emily Diamond, Melva Treviño Peña, Jesse Reiblich, Elizabeth Mendenhall, Travess Smalley, Kendall Moore, Amber Neville, Jaclyn Witterschein, Emi Uchida, Soni Pradhanang, and many others. Taûbotne anawáyean (thank you) to Lorén Spears at the Tomaquag Museum and Maryann Mathews at the Manissean Tribal Council for their years of collaboration with the DWELL Lab, as well as to Ryan Kopp and Rebecca Reeves at the Stormwater Innovation Center and our many other community partners.

Thank you to the numerous undergraduate students who have taken my courses on science writing and digital rhetoric, the graduate students who participated in my courses on the rhetoric of science, science communication, and environmental advocacy, and to the many students who have collaborated as researchers in the DWELL Lab, especially Hannah MacDonald, Ellen Fritz, Gabrielle Pezich, Abbey Greene, Julian Garrison, Ally Overbay, AnnaFaith Jorgensen, Ally Cuomo, Christina DiCenzo, Ashley Katusa, Joseph Ahart, Sarah O'Sullivan, Aidan Donnellan, Calder Puckett, Novena Kapisa, Asta Habtemichael, Zaria Griffith, Erica Meier, Ellen Fritz, and Erin Edmonds, and especially to research assistants Fran Webber and Taylor Roberts, who both provided valuable edits and corrections for this manuscript. I would also like to thank my fellow EmerAgency collaborators Gregory L. Ulmer, John Craig Freeman, and Jack Stenner for sustaining the intellectual energy that goes into writing a book and for always encouraging and supporting me. Though Ulmer retired at the onset of my time at UF, we often met for coffee, and I learned more from those conversations on the bench outside Little Hall than I have in any seminar. Thank you to Sergio Figueiredo, Lydia Ferguson, Kenneth Walker, Donnie Sackey, Tim Amidon, Bridie McGreavy, David Grant, Byron Hawk, Jennifer Clary-Lemon, Joshua Trey Barnett, Nate Johnson, Lydia Wilkes, Crystal Colombini, Derek Ross, Chad Wickman, Randall Monty, Casey Boyle, Charles Woods, Morgan Banville, Lauren Cagle, Ira Allen, Ralph Cintron, Thomas Rickert, Johanna Barthmaier-Payne, and so many others for conversation, guidance, and inspiration along the way. I count myself quite lucky to get to collaborate with such incredible colleagues, mentors, peers, and students.

In addition to this intellectual support, I have been fortunate to receive financial support that made this project possible. I am grateful to the URI College of Arts & Sciences and College of Environmental and Life Sciences, who provided me with release time from teaching to focus on writing, as well as travel support for the research that made this book possible. An important part of researching this book was traveling to the locations where some (but not nearly all) of the Odum brothers' research took place and interviewing researchers who had worked with the Odums and who were conducting contemporary research at sites they established. Gracious thanks to the scientists who took time out of their busy schedules to speak with me. This study was approved by the URI Institutional Review Board (IRB Protocol Number: 2232826-1). I would like to thank NIFA/RIAES (Grant Number: NE1962) and the URI Harrington School of Communication and Media for generously providing me with travel support to visit the Eugene P. Odum Archives at the University of Georgia (UGA) and the URI Center for Humanities for providing me with a faculty research grant to travel to the Howard T. Odum Collection at the University of Florida and for subvention funding to support open access. Special thanks to John Nemmers, Florence Turcotte, Caleb Del Rio, and Michele Wilbanks at the George A. Smathers Library at UF and Mazie Bowen at UGA's Hargrett Special Collections Library for their assistance. Ideas and findings from this book have been presented at conferences for the National Communication Association, Rhetoric Society of America, and College Composition and Communication, and benefited from audiences at the University of Nevada and as part of the performance lecture *Double Counting: The Odum Oration* by Jamie Allen and Karolina Sobecka at the 2020 transmediale festival at the Berlin Volksbühne. Portions of this book have appeared in the form of an article, "A Counterhistory of Rhetorical Ecologies," published in *Rhetoric Society Quarterly*, and benefited enormously from the peer review process. I would like to thank the journal editor Jacqueline Rhodes, and editorial assistant Rebecca Conklin, as well as the three anonymous reviewers, whose generous feedback helped refine and develop this project through multiple revisions. My gratitude goes out to the editor Robin James and anonymous reviewers at Palgrave who helped me work through the process of compiling and revising the book.

Special and most gracious thanks are due to Lee Rozelle, professor of English at the University of Montevallo, without whose decades of exceptional mentorship, pure comedy, and sage advice I would not be here today (one could say that this is all your fault). Alongside Lee, Adam Roberts and Kimberly Wright—and the other occasional members of the Smokey Hollow Writers Workshop—have also read more than their fair share of my writing over the many summers we've spent together eating bibimbap across Alabama. I owe additional appreciation to Aaron Beveridge, Jacob Greene, and Charlie Sterchi, who each offered helpful feedback, edits, and advice along the path.

And most of all I thank my family, both for giving me space and time to write and for the welcome and needed distractions from writing. I am most grateful for your company on some of my research travels and for your patience and resilience when we experienced island-wide rolling blackouts in Puerto Rico. This book is dedicated to you. To my wife, Jane, a brilliant educator and patient interlocutor, who has entertained far more than her allotment of discussions around the ideas in this book: "*lo que la primavera hace con los cerezos.*" And to my children, Cate and Madison: You remind me every day that the world is filled with beauty and wonder if you look at it right.

Competing Interests The author has no competing interests to declare that are relevant to the content of this manuscript.

Contents

1 **Ecology, Rhetoric, and Complex Systems**   1
   Inventing Ecologies   3
   Technomorphism in the Rhetoric of Ecosystems   4
   The Machine Starts?   8
   The Use and Abuse of Vegetational Concepts and Terms   11
   Field Histories   13
   Rhetoric's Ecological Moment   19
   Progression of the Book   26

2 **Energy Rhetoric in Odum's "Silver Springs Study"**   45
   Silver Springs and the Ecosystem   46
   Energy Systems   50
   Energy and/as Rhetoric   52
   Circulation, Emergy, and the Rheme   54
   Ecosystem as Diagram and Apparatus   59
   Energy and Consciousness   61

3 **The Trouble with Ecology: The Afterlives of Radioecology in the Marshall Islands and Puerto Rico**   71
   The Trouble with Ecology   73
   Colonial Ecologies of Nuclear Pollution   81
   The Rhetoric of Science Meets the Rhetoric of Technology   83
   Islands in the Scheme   84

4 Ecology Out of Place: *Topoi* and Spatial Problems ... 91
   Locating Ecosystems ... 92
   Scale and its Discontents ... 94
   Scale Critique ... 100
   Scale and Ecosystems Ecology ... 102
   The Machine Flops? ... 105
   Whitey on the Moon ... 107

5 Ecology Out of Time: *Kairos* and Temporal Problems ... 115
   The Problem of Time ... 117
   Kairos and the Ecosystem ... 120
   From Anthropocene to Technocene Rhetorics ... 123
   Kairotic Ecologies ... 125

6 Coda: Feedback Loops ... 133
   Tailing Rhetorical Ecologies ... 135
   Field Notes from the DWELL Lab ... 138

Index ... 147

About the Author

Madison P. Jones is an associate professor of Professional/Public Writing and Natural Resources Science at the University of Rhode Island where he is a senior fellow at the Coastal Institute, coordinates the Science Writing & Rhetoric graduate certificate program, and teaches courses on science writing, environmental rhetoric, and public advocacy. He is the founding director of the Digital Writing Environments, Location, and Localization (DWELL) Lab, a digital humanities initiative that builds community-engaged media projects through co-creation and collaboration. To learn more about DWELL, visit: uri.edu/dwell.

Jones has conducted research everywhere from Athens, Georgia, to Athens, Greece. His award-winning scholarship intersects the environmental and digital humanities and has been supported by grants from the National Science Foundation, the National Endowment for the Humanities, and the National Institute of Food and Agriculture. His recent work has been recognized by awards from the American Academy of Environmental Engineers and Scientists, the Society for Literature, Science, and the Arts, and other organizations. Jones is author of over twenty articles and chapters and several books on the environmental humanities, including *Rhetorical Ecologies* (NCTE, 2024) and *Writing the Environment in Nineteenth-Century American Literature* (Bloomsbury, 2015). He is also a poet and author of *Losing the Dog* (Salmon, forthcoming) and *Reflections on the Dark Water* (Solomon & George, 2016). For more information, visit: madisonpjones.com.

Abbreviations

AEC	US Atomic Energy Commission
ANT	Actor-network theory
CLEAR	Civic Laboratory for Environmental Action Research
DH	Digital humanities
DWELL Lab	Digital Writing Environments, Location, and Localization Lab
ESA	Ecological Society of America
EH	Environmental humanities
HCI	Human–computer interaction
LTER	Long-Term Ecological Research
NSF	National Science Foundation
RNM	Rhetorical new materialisms
RSTM	Rhetoric of science, technology, and medicine
RWCS	Rhetoric, writing, and communication studies
SREL	Savannah River Ecology Laboratory
STS	Science and technology studies

List of Figures

Fig. 1.1   Aerial image of the July 1, 1946, Able detonation at the Bikini Atoll (United States Army Air Forces), the first detonation of Operation Crossroads (consisting of Able, Baker, and Charlie). Retrieved from Wikimedia Commons, https://commons.wikimedia.org/wiki/File:Operation_Crossroads_-_Able_001.jpg (Public Domain)   23

Fig. 2.1   Sample visualizations of H. T. Odum's "System of Generic Symbols," adapted with permission from Sholto Maud's stencils, via Wikimedia Commons, https://commons.wikimedia.org/wiki/File:Energy_Systems_Symbols_H.gif (Public Domain). This iconographic approach to ecosystem modeling was introduced by H. T. Odum and Elisabeth C. Odum in *Modeling for All Scales: An Introduction to System Simulation* (2000)   51

Fig. 2.2   An example of H. T. Odum's energy diagrams, depicting conflict in Afghanistan and the relationships between energy, war, religion, ideology, and media representation. Courtesy of Special & Area Studies Collections at the University of Florida   55

Fig. 2.3   Ad promoting the 1975 Energy and Consciousness event at the University of Florida that featured H. T. Odum and beat poets (University of Florida 1975)   62

Fig. 3.1   "Source and Crew. El Verde. June 1965." The crew at El Verde, led by Howard T. Odum, with a cylinder that contains a 10,000-curie cesium gamma radiation source. The source was deposited in the forest by helicopter in 1964 as part of a radioecology study that took place from 1962 until 1970 and

	was funded by the AEC. Courtesy of the Howard T. Odum Papers, Special and Area Studies Collections, George A. Smathers Libraries, University of Florida, Gainesville, Florida	75
Fig. 3.2	An aerial photograph of Runit Dome (sometimes referred to as Cactus crater containment structure) on Runit Island, Enewetak Atoll. The crater that was created by the United States during the detonations of Operation Hardtack I in 1958 was used to create a burial pit to cover the radioactive material that was scraped from the contaminated Enewetak Atoll islands. Courtesy of Wikimedia Commons, https://commons.wikimedia.org/wiki/File:Runit_Dome_001.jpg (Public Domain)	76
Fig. 4.1	A drawing of the *scala naturae*, or the Great Chain of Being, by Fray Diego de Valadés, *Rhetorica Christiana* (1597). Wikimedia Commons. Retrieved from https://commons.wikimedia.org/w/index.php?curid=33603873	95
Fig. 4.2	An 1837 sketch by Charles Darwin (Fig. 4.2A) from his *First Notebook on Transmutation of Species*, visualizing an evolutionary tree, likely his first diagram representing evolution, courtesy of Wikimedia Commons. Retrieved from https://commons.wikimedia.org/wiki/File:Darwin_tree.png.A similar tree-like structure (Fig. 4.2B) was used by Ernst Haeckel in his 1866 book *General Morphology of Organisms* to visualize what he dubbed "phylogeny," or the evolutionary history of an organism, courtesy of Wikimedia Commons. Retrieved from https://commons.wikimedia.org/wiki/File:Haeckel_arbol_bn.png. His "Pedigree of Man" visual (Fig. 4.2C), published in his 1879 book *The Evolution of Man*, most clearly reproduces the *scala naturae*, courtesy of Wikimedia Commons. Retrieved from https://en.wikipedia.org/wiki/File:Tree_of_life_by_Haeckel.jpg	97
Fig. 4.3	A conceptual model used to describe the structure and dynamics of an ecosystem, based on H. T. Odum's Silver Springs study, courtesy of *Biology 2e* from OpenStax, licensed under Creative Commons Attribution License v4.0. Importantly, energy decreases with each order of magnitude	98
Fig. 4.4	"Silver Springs Model" from Odum's *Environment, Power and Society* (1971), courtesy of Wikimedia, https://commons.wikimedia.org/wiki/File:Silver_Spring_Model.jpg. The model represents herbivores, carnivores, and decomposers, as well as the flow of energy through the system	103
Fig. 4.5	Illustration of the macroscopic perspective from Odum's *Environment, Power and Society* (p. 10)	104

CHAPTER 1

Ecology, Rhetoric, and Complex Systems

Abstract Building from scholarship in the rhetoric of science, technology, and medicine and rhetorical ecologies, this chapter investigates how ecosystems ecology was influenced by technological metaphors that serve to both structure and constrain ecological inquiry. Turning to the intellectual history of the ecosystem as a concept and a diagram, this chapter moves from its origins in the work of Arthur Tansley to its development by Howard T. and Eugene P. Odum. This context demonstrates how the legacy of technological thinking shapes contemporary environmental inquiry. Through interdisciplinary research, this chapter analyzes ecosystems as diagrammatic scientific models and rhetorical inventions.

Keywords Rhetoric of science and technology • Ecosystems ecology • Cybernetics • Systems theory • Environmental humanities • Technomorphism

> Of course she knew all about the communication-system. There was nothing mysterious in it [...] the system had been in use for many, many years, long before the universal establishment of the Machine.—E. M. Forster, "The Machine Stops"[1]

University of Georgia, Athens, Georgia, United States
I glance down at my phone and make a back-of-the-envelope calculation, comparing the twenty-minute walk back to my hotel room to the coming weather mass on the radar map. The encroaching signs of Hurricane Helene are everywhere around me. The morning sounds of birdsong and cicada chorus that greeted me on my walk here have been replaced with the hiss of trees bending in the wind. I pick up the pace and wonder if I will feel up to returning to the Special Collections Library when they reopen after lunch. Yesterday afternoon, I made the decision to cut my trip short, as the spaghetti models of the storm's path wove a tighter weft of certainty, and predictions of its scale and force grew more dire. All week, I had been tracking Helene's path, first seeing reports of the possible tropical storm during a layover on the trip down and then watching with growing anxiety as the storm met with the Gulf of Mexico's extraordinarily and unseasonably high water temperatures, increasing rapidly in velocity and magnitude. I had spent all morning hurriedly working in the stately silence of the Special Collections Library's grand reading room, furiously photographing documents and triaging the remaining time I would have to go through more than a dozen unexamined boxes I had pulled from the Eugene P. Odum archival collection. Out the window, I watched as the sky turned and the trees began to bend in the breeze. While I had hoped for another day with the rich materials, this trip had already provided me with an incredible opportunity to look through a window into the life of Eugene Odum, one of the figures who had loomed large in my research on ecology over the better part of the last decade. I first found the trail of Eugene through his younger brother Howard T. Odum, who had a distinguished career at the University of Florida where I completed my doctorate. At the University of Georgia, Eugene founded the Savannah River Ecology Laboratory (SREL) and the Institute of Ecology, which grew into the Odum School of Ecology. Together, the Odum brothers had left a lasting mark on the field of ecology, and as was evident in the hundreds of letters I had read over the past few days, touched the lives of so many different people. I had spent many hours following the course of Eugene's vast body of work, from its beginnings with a modest grant of $10,000 from the Atomic Energy Commission to leading massive-scale ecological research projects all over the world. My mind races with the many new things I had learned about the Odum brothers' work, deepening and challenging my understanding of the ways that they had imagined, invented, and ultimately revolutionized modern ecological thought. Sudden rainfall interrupts these ruminations as my thoughts turn toward shelter.

Inventing Ecologies

While today it is virtually impossible to imagine our relationship with the so-called nonhuman world without the structuring metaphors and analogies of environment, ecology, and the ecosystem, these concepts each emerged from a dynamic set of historical moments. For instance, the neologism "environment" was coined by the historian Thomas Carlyle's 1828 loose translation (or arguably even mistranslation) of the word "*Umgebung*" in the work of Johann Wolfgang von Goethe.[2] As a concept, *environment* opened radical possibilities for scientists to theorize and study the relations between biotic and abiotic communities.[3] Together with *Umwelt* and *Innenwelt*, *Umgebung* was also taken up in semiotics and communication theory by Jakob von Uexküll's 1920 *Theoretische Biologie*, linking biosphere and semiosphere. The subsequent invention of the concepts of *ecology* by Ernst Haeckel in 1866[4]—building from Charles Darwin's metaphor of an "economy of nature"—and then the *ecosystem*—which was introduced to the field by Arthur Tansley in 1935—together brought about what Thomas Kuhn terms a "paradigm shift" through successive rhetorical moves away from empirical naturalism toward theoretical, qualitative, and then quantitative approaches to studying environments.[5] The ecosystem has its own unique conceptual history. Tansley published the first paper using the term in 1935, "The Use and Abuse of Vegetational Concepts and Terms," criticizing the common usage of terms like "climax" and "succession" as deterministic and introducing ecosystems as a better way to describe the dynamic relationships between biotic and abiotic entities within the conditions of a community.[6] Tansley drew upon metaphors from mathematics, economics, and systems theory, as well as inspiration from Sigmund Freud's idea that the human brain comprised an electrical network.[7] Thus, the ecosystem story begins with an electrical spark of invention, leading to innovations that would activate an entire disciplinary field, actualize a new approach to ecology, and catalyze experimental systems research across a wide range of disciplines. However, the term itself may have been suggested to Tansley by his colleague, Arthur Roy Clapham, an element of its conceptual ecology that is often left out of the ecosystem story.[8] [9] From this origin point, the ecosystem concept becomes further entangled with Tansley's interests and involvement in socialism, Freudian psychology, and British imperialism.[10]

This theoretical concept was then put into practice by Raymond Lindeman in a 1942 study mapping the trophic dynamics of Cedar Bog

Lake, Minnesota. The 1953 publication of the textbook *Fundamentals of Ecology*, written by Eugene Odum in collaboration with his younger brother Howard T. Odum, was the field's only textbook for over a decade, further cementing the ecosystem as a central concept for ecological science. As historian Benjamin Golley points out, "…the ecosystem story is largely an American tale" that began to rapidly unfold following WWII as "Europe and Japan were preoccupied with reconstruction" and scientists worldwide "repudiated aspects of ecological theory that had been used by the Nazis [...] to force conformity on the population and to base racist policy."[11] Such applications did not happen in a cultural vacuum, as Haeckel himself was a proponent of Social Darwinism, eugenics, and scientific racism, as was Uexküll, whose "*Umwelt* theory is antidemocratic, totalitarian, and holistic in the worst sense," and further, he was "deeply involved in Nazism."[12] Likewise, as Thomas Patrick Pringle explains, Tansley regarded the ecosystem as directly opposing and conflicting with "the idea of holism propounded by Jan Smuts, the South African political leader, general, ecologist, and one of the philosophical architects of apartheid."[13] Smuts' *Holism and Evolution* opens with an epigraph from Plato's *Sophist*: "That which comes to be always does so as a whole; so that if a man does not count the whole among realities he ought not to speak of substance or of coming-to-be as real."[14] In two separate places in the book, Smuts directly references Plato as he introduces and defines the concept of holism, suggesting further connections between rhetoric, philosophy, and science at the core of ecological theories of holism.[15] While theoretically, ecology presented dubious ideological grounds—with obvious ties to German philosophers and scientists like Goethe, Uexküll, and Haeckel—ecosystems offered an appealing model of modern American science, one grounded in the rhetoric of purity through data-driven methodologies, abstract models, and cutting-edge computer technologies.[16] At the same time, the ecosystem introduced its own set of fraught rhetorical complexities.

Technomorphism in the Rhetoric of Ecosystems

The need for this book arises from the perennial rhetorical problems that have plagued ecology since its inception, and which reached something of a fever pitch with the crystallization of ecosystems ecology during the mid-to-late twentieth century. While critiques of anthropomorphism/anthropocentrism are a staple of the environmental humanities (EH), this book is interested in the ways that *technomorphism*—a term that describes the rhetorical application of machine characteristics to other phenomena (such as humans, plants, animals, chemicals, microbes, processes, and the environment)—both shapes and limits ecological inquiry, obscuring some of ecology's problems through the rhetoric of technology.[17] The concept functions similarly to anthropomorphism, which ascribes human features to nonhuman phenomena. Technomorphism has generally only been applied by scholars to human contexts, especially in human–computer interaction (HCI);[18] this book extends the concept to show how ecosystem researchers apply technological and mechanistic metaphors to characterize environments and even life itself as machines all the way down.[19] In doing so, it brings together perspectives informed by scholarship in the rhetoric of science, technology, and medicine (RSTM) in order to examine the roles that technology increasingly plays in shaping ecological inquiry across the production of both rhetorical and scientific knowledge. As such, *Inventing Ecosystems* focuses on the ways that technology participates in co-producing the so-called Anthropocene (a topic that we will return to in Chap. 4), arguing that we are living as much in the *Technocene*, a time where technology is drastically reshaping the relations that constitute humans, HCI, and the environments with which we interact.

As the ecosystem concept emerged and was solidified in the work of Eugene and H. T. Odum and many others, it carried with it many dubious underlying rhetorical assumptions that drastically shaped, and in many ways limited, ecological inquiry. In recent years, the fields of rhetoric, writing, and communication studies (RWCS)—alongside other fields in EH, digital humanities (DH), and across the social sciences—have enthusiastically, and with an increasing velocity, taken up ecology, networks, circulation, and systems as central animating metaphors for the field, as well as producing numerous studies of environmentalist discourse, writing ecologies, and environmental rhetoric. As big data and artificial intelligence (AI) have garnered massive widespread popularity following OpenAI's release of ChatGPT in 2022, the field's surging interest in AI has spread well beyond a focus on DH, machine learning, and computational rhetoric. Meanwhile, our social networks have become persuasive environments with HCI directed algorithmically.[20] With the growing popularity of persuasive technologies, the use of both anthropomorphic and technomorphic rhetoric has likewise increased.[21] However, such hasty moves to borrow metaphors from science and technology risks importing some of the major problems that have long troubled disciplines like ecology, from problems of analogy, metaphor, and scale to technological determinism, extractivism, frontierism, and solutionism. This book bridges these interdisciplinary discussions, critically examining how the adoption of technological metaphors in ecology has shaped not only scientific practices but also the ways that RWCS engages with ecological inquiry. The concept of invention has longstanding ties to rhetoric, framed by Aristotle as a primary mechanism of persuasion itself, and codified by Cicero as one of the five canons of rhetoric. Similarly, invention has deep roots in RSTM, where innovation and discovery are driving forces.[22] Important early works like Lawrence J. Prelli's *A Rhetoric of Science* frame invention as a (perhaps even *the*) key concern for interpreting science through and with rhetorical perspectives.[23] This book partly examines the political, social, and ethical impacts of the application of technological and mechanical metaphors to ecology as a form of technomorphic rhetorical invention.

Inventing Ecosystems argues that our field's long engagement with RSTM,[24] rhetorical new materialisms (RNM),[25] and ecocomposition[26] has well equipped scholars to both interrogate and critique the inherited theoretical problems of ecology, as well as to develop new methods and practices that offer more just and nuanced approaches to ecological inquiry and more closely follow the lead of contemporary ecological science. While RSTM scholars have traditionally distinguished between the rhetoric of science and the rhetoric of technology as distinct foci,[27] the technomorphic rhetoric of ecosystems ecology makes requisite a combined approach to the rhetoric of science and technology. By examining the role of energy, cybernetics, and colonial histories in shaping ecosystems ecology, this book highlights the critical role that rhetoric plays in understanding how science interacts with the natural world as well as the social and historical dimensions of science. This interdisciplinary approach is urgently needed, as ecological science today is increasingly called upon to address complex environmental challenges like climate change, biodiversity loss, and habitat destruction—issues that are deeply entangled with global histories of colonialism and exploitation, what Ira Allen refers to as "CaCaCo," describing the "polycrisis" of "the carbon-capitalism-colonialism assemblage."[28] While the last decade has seen a dramatic uptick in RWCS scholarship focusing on the environmental crisis[29] and its impacts on how we persuade and communicate,[30] as well as studies of rhetoric in geology and geoengineering,[31] rhetoric and ecology[32] rhetoric and climatology,[33] and discourse analysis in fields like paleontology[34] and microbiology,[35] fewer works place our contemporary moment within the context of environmental history and rhetorical historiography. By focusing on the rhetorical dimensions of ecosystems research, this book aims to illuminate the ways that scientific practices have been shaped by colonial ideologies and to propose alternatives that foreground community-based science communication and anti-colonial methodologies.[36] Ultimately, this book not only reveals the rhetorical foundations of ecology but also offers practical strategies for improving both the theory and practice of rhetorical ecological inquiry.

As I will return to throughout this book, the shift to ecosystems revolutionized the capabilities of ecologists to better account for vast complexities, spatial scales, and time periods. The broad moves withing ecological science from abstract holism to more reductive, relativistic, and mechanistic approaches to the field brought about more theoretically subdued but methodologically and practically innovative studies. The eco*system* metaphor also subverted Carlyle's radical anti-mechanical rhetoric of environment even as it provided methods to study complex ecological structures.[37] Such a shifting rhetorical landscape presents a clear need for RWCS scholars interested in ecology and the environment to come to terms with longstanding conversations in media studies, ecocriticism, the history and philosophy of science, and environmental studies. At the same time, I argue throughout this book that ecology and rhetoric have much to gain from each other. This project builds from Dana Phillips' important ecocritical work *The Truth of Ecology*, where he argues that, for ecology, "…an over-reliance on analogy and metaphor has posed an obstacle to the advance of theory and research. That it must struggle with rhetorical issues would seem to link ecology's misfortunes with troubles of a sort familiar to students of the humanities."[38] The ecosystem conceptually exacerbates the problems of analogy and metaphor and increases ecology's rhetorical misfortunes by orders of magnitude. These complexities clearly necessitate interdisciplinary interventions that work between, within, and among disciplines in both the humanities and sciences to address these problems and develop better theories, methods, and practices for rhetorical ecological inquiry. This book examines the invention and development of the ecosystem through perspectives informed by rhetoric and historiography in order to reframe some of the intractable problems that ecosystems have presented to ecological thinking in hopes of finding novel solutions.

The shift from the "old" science of ecology, which evolved out of natural history, to the "new" ecosystems ecology likewise invokes what Erich Hörl refers to as "the nature / technics divide" in ecology, describing how these different concepts pivot between centering the technological or the natural world in conceptions of phenomena like agency and relationality.[39] The technoecological conditions of ecology become an oscillating teleology between technosphere and biosphere. As Pringle puts it, this divide describes the way that "environments have been conceived as media technologies and vice versa."[40] Peter Taylor has demonstrated how early ecosystems ecologists capitalized on these metaphors by promoting a "technocratic optimism," which built from the technocracy movement of the 1930s that promised to replace the chaotic landscape of economics, politics, and environmental disaster following the Great Depression with a system based on the science of energy use.[41] From the Ancient Greek word *techne* (τέχνη), meaning roughly "craft" or "art," and *kratos*, meaning political power, regime, or rule, technocracy sought to solve the world's problems by establishing scientists and technologists as the elite rulers of society. Howard Scott, who founded the technocracy movement, viewed the choice to replace democracy with rule by technocrats as deciding between "science or chaos."[42] Scott viewed this decision as inevitable, using the logic of technological determinism to build his case.[43] While the technocracy movement was short-lived, it persisted in shaping the earliest theories of ecosystems ecology, which in turn continues to impact environmental thought today.[44] Hörl, referring to the theories of Gilles Deleuze and Félix Guattari, explains how this process leads to a "technoecological condition" that, on the one hand, suggests that we are contained (environed) by a mechanistic system, and on the other, effaces the rhetorical moves that lead to "the apparatus of capture of Environmentality."[45] In other words, as ecologists increasingly began to borrow from systems, economic, and cybernetic theories in the 1940s and 1950s to understand the natural world, they began to naturalize the technomorphic and technocratic idea that environments were mechanistic, and as such, were best understood with sensors and algorithms, and regulated (or controlled) as (and through) machines.

The Machine Starts?

Written over a hundred years before our current anxieties about the technocratic optimism associated with public figures like Elon Musk, Jeff Bezos, and Sam Altman, the rise of AI and other emerging networked technologies, and the disastrous effects that these technologies have had on global social, political, and environmental domains, E. M. Forster's

"The Machine Stops" offers a nuanced take on the encroachment of technology on the natural world.[46] Printed more than twenty-five years before Tansley published the first paper using the term "ecosystem," Forster's peculiar and prescient short story—which appeared in *The Oxford and Cambridge Review* in 1909—provides a fable that places the rhetorical conflicts between technosphere and biosphere into direct narrative tension. Haeckel and ecology were in the air in England in the 1890s, with his work becoming more popular there than in Germany at the time.[47] "The Machine Stops" offers a useful glimpse at the roots of our current zeitgeist, as ecology emerged alongside modernism as part of what E.H. Howell has referred to as a "Modernist Anthropocene."[48] In the story, the surface of Earth has been rendered uninhabitable for animal life by an atmosphere that is believed to be poisonous to everything but small plants. Deep below the surface of the planet, humans live in massive subterranean cities that are referred to as "the Machine." Like a colossal life-support apparatus, this mechanistic world provides everything that humans need to survive in their individual, isolated hive-like pods, where they communicate over a network that resembles the Internet and through screens that bear an uncanny resemblance to both television and modern video conferencing. Physical contact, both between people and with the outside world, is verboten, and the subterranean denizens instead worship abstract knowledge and isolation. Forster does not use the words "environment" or "ecology," which had not achieved widespread usage at the time, but rather, he continually refers to the organization of humans and their technosphere as an "apparatus" and a "system"—be it regarding telepresence for communications, conveyors for moving people and products, or the religious practices surrounding worship of the Machine.

The story focuses on two major characters: Vashti, a renowned lecturer who spends her days pondering secondhand ideas, and Kuno, her son, who lives on the other side of the world and who is a subversive rebel that rejects the ideology of the Machine. While relatively few scholars have discussed this extraordinary story, it has sometimes been characterized as an anomaly in Forster's body of work.[49] However, the novel clearly expresses modernist anxieties about emerging technology that can be seen throughout Forster's body of work, such as in his masterpiece *Howard's End*, which opens with the words "Only Connect,"[50] and was published the following year after "The Machine Stops."[51] In *Howard's End*, upper-class idealization of the pastoral is contrasted with encroachment of industrial society of modern London on the rural countryside. Along similar lines, Kuno seeks to return to a "primitive" way of living in the biosphere that exists outside the bounds of the Machine. Vashti, on the other hand, is a devotee of the technosphere who sees no reason to reach beyond the telepresence technologies provided by the Machine. While Alf Seegert argues that the story "succumbs to a modernist nostalgia for something that never existed,"[52] I view "The Machine Stops" as expressing a more profound ambivalence toward the conflation of technosphere and biosphere in the modernist project, characterized by the divergence of Kuno's pastoral idealism (à la Leo Marx) and Vashti's technocratic optimism. The story oscillates between dystopian and utopian visions of technosphere and biosphere. In this cautionary—and reactionary—tale neither character is ultimately spared from the apocalyptic fate of the Machine, suggesting the ways that technomorphism produces ideological constraints.

Forster's "The Machine Stops" is a prophetic text, imagining coming tensions between ecological and technological systems that are now central to contemporary environmental thought. If ecology and modernism share overlapping problems and concerns that mark the advent of the twentieth century, the ecosystem presents knowledge problems that persist today in our networked publics—and in a "nature" that Phillips defines through the hyperreal of postmodernism[53]—as systems thinking has become central to the study of both the natural world and the technosphere.[54] The tensions between technocratic control and chaos are evident in the ways that both ecosystems and machines are understood through rhetoric and how they shape the political and environmental landscape. This book is grounded in the understanding that the ecosystem concept is not only a scientific model but also a rhetorical tool that shapes our understanding of both natural and technological systems. Much like Forster's "Machine," the ecosystem concept has developed in the context of technological advances and is intertwined with the rhetoric of cybernetics and technocratic control that emerged in the mid-twentieth century. By tracing the history of the ecosystem concept through the work of Eugene and Howard T. Odum, as well as examining its relationship with colonial practices in places like the Marshall Islands and Puerto Rico, this book uncovers the deeply embedded assumptions about power, control, and exploitation that have long shaped ecological research. In "The Machine Stops," the technosphere collapses, killing all who rely upon it. Today, anthropogenic environmental changes pose similar existential threats to complex processes that sustain life on earth. The technological metaphors that animate ecosystems can serve to naturalize technocratic perspectives, furthering the intractable nature of these problems. Instead, *Inventing Ecosystems* challenges environmental inquiry to seek more fragmented and chastened ways of engaging with the natural world.

THE USE AND ABUSE OF VEGETATIONAL CONCEPTS AND TERMS

In Adam Curtis' 2011 "The Use and Abuse of Vegetational Concepts" (named after Tansley's iconic paper), the second episode in the television series *All Watched Over by Machines of Loving Grace*, he discusses the invention and evolution of the ecosystem concept.[55] He examines how cybernetics and systems theories contributed to a mechanistic understanding of nature as a self-regulating system. Tansley built from Jay Forrester's work with feedback loops, focusing on the ways that complex systems are thought to work toward balance, stability, and equilibrium. These theories were taken up by Eugene and H. T. Odum's research, which positioned humans, machines, and ecosystems as nodes within a self-regulating cybernetic network. These ideas inspired cultural movements like Buckminster Fuller's radical architecture (discussed in Chap. 4) and Richard Brautigan's poetry (for whose eponymous book Curtis' series is named), which imagined the fusion of biosphere and technosphere. Brautigan's titular poem imagines a utopian balance between humans and the natural world, restored by "cybernetic ecology," exclaiming, "I like to think / (right now, please!) / of a cybernetic forest / filled with pines and electronics / where deer stroll peacefully / past computers / as if they were flowers / with spinning blossoms."[56] Such a vision supports an optimism that sees technocratic control as a necessary component of a sustainable future. Cybernetic ecology, as Brautigan presents it, offers a new way to return to a pastoral vision of nature. However, Curtis demonstrates that these mechanistic models began to disintegrate when they were applied to real-world studies, which demonstrated that natural systems are far less stable than ecosystems ecology originally supposed. Curtis' exploration of the development and disintegration of mechanistic models of ecosystems parallels the ways that digital networks similarly rely on material environments, revealing the interdependence of technological systems and environmental conditions.

Just as the ecosystem concept rhetorically transforms the natural world into a cybernetic network, so do digital networks themselves rely on environments to sustain them. In her pathbreaking book, *The Undersea Network*, Nicole Starosielski follows the undersea telecommunication cable system which today carries nearly all transoceanic Internet traffic, tracing these cable systems as they traverse the various companies, regions, eras, environments, borders, and cultures they move through to span a globe.[57] While we tend to imagine communication networks as clouds, Starosielski demonstrates how those networks rely materially on specific places, peoples, and environments to make signal transmission possible. Her project inverts the previous work of Neal Stephenson's iconic article, "Mother Earth Mother Board," which ultimately views these environments as *constraints* for cable systems.[58] Instead, Starosielski demonstrates that environments, places, people, and histories are the very *conditions* through which the networks are made possible. In order to span the globe, these cables must surface at various points, places which Starosielski refers to as "networked islands."[59] Many of these island sites were also crucial locations for the pioneering research being carried out by H. T. and Eugene Odum, such as in Puerto Rico and the Marshall Islands. This context becomes important for understanding ecosystems, because, while "islands and networks appear to be mutually exclusive,"[60] she points out how the word for island actually indicates "isolating and insulating" and that "isolation and boundedness are what make [geographical islands] special."[61] Thus, these island spaces are threatened by the emergence of the network cables in the "cultural anxiety that islandness will disappear."[62] As I will discuss in greater length in Chap. 3, just as islands help triangulate the global network, so does their boundedness and insulation provide the very conditions that make the ecosystem concept possible. These "isolated" and bounded environments provided a particular place that could manifest and support the ecosystem concept.

Today, ecology finds conceptual purchase in numerous contexts outside of the environmental sciences. In the introduction to our recent edited collection, *Rhetorical Ecologies*, Sid Dobrin and I discuss the "profound and pervasive cultural and multidisciplinary influence" of ecology.[63] We follow the work of Félix Guattari, who discusses how the concept mushroomed in his 1989 book *The Three Ecologies* and applies the term to describe the social, mental, and environmental aspects of his theory of "ecosophy" in order "to comprehend the interactions between ecosystems [and] the mechanosphere."[64] Along similar lines, Deleuze and Guattari's previous book *A Thousand Plateaus: Capitalism and Schizophrenia* helped solidify ecology as an influential philosophical concept beyond only its original scientific context.[65] Eugene Odum himself explored the link between social science and ecology, comparing the ecosystem to "cultural units functioning together as a whole."[66] Across the arts, humanities, and social sciences over the past few decades there have been a variety of "ecological turns," denoting a wide variety of applications of ecologies as a concept within different fields to understand everything from nature to digital networks.[67] The results vary widely, but they are often inconsistently applied, pedantic, and ambiguous.[68] For example, Pringle[69] demonstrates how Smuts'[70] ecological holism and racist apartheid philosophy persist in shaping contemporary ecological inquiry.

As such, this book concerns itself rhetorically with ecology broadly, and more specifically the ecosystem, less as an analogy or metaphor, and more as a *diagram* in the sense that Deleuze and Guattari deploy the term. The function of the diagram is more than merely representing or modeling a phenomenon. Rather, diagrams actively participate in the processes and practices of meaning-making. In this book, I bring recent approaches to diagrammatic rhetoric[71] [72] together with media studies scholars who view the "ecosystem as apparatus"[73] in order to understand the diagrammatic rhetoric of ecosystems. Along similar lines, this book seeks to unearth a history of the ecosystem concept, following the conceptual histories Raymond Williams engages in *Culture and Society*[74] and *Keywords*.[75] Williams describes a need to carefully interrogate the ways that concepts serve as ideological screens that filter our experiences. While the ecosystem has played a pivotal role in shaping our contemporary ecological imagination, it is often overlooked in keyword projects in the environmental humanities (EH).[76] With this in mind, this book aims to foreground the role that the ecosystem concept played in shaping modern ecological thought across the sciences, arts, humanities, and social sciences.

Field Histories

Thus far, I have focused on tracing the conceptual and material connections between ecology and the rhetoric of technocratic optimism. In what follows, I will examine its relationship to the nuclear colonial violence of the Anthropocene, and I will detail how the spatiotemporal imbroglio of rhetorical ecologies is rooted in these connections. While this discussion of the limitations and problems of ecosystems ecology may lead some to conclude that RWCS scholars (and perhaps even scientists themselves) should simply replace the concept with another term, I argue instead that field histories allow scholars to (re)place such concepts back into a rhetorical-historiographic framework. In following the conceptual history of the ecosystem, this book embodies what I refer to elsewhere as *field histories*, a methodology for tracing and understanding interdisciplinary relationships within, between, and among different fields.[77] Field histories are the study of disciplinary conceptual histories as they emerge, evolve, overlap, intersect, diverge, and influence inquiry. As rhetoricians interested in ecology are participating in the social justice paradigm taking place across RWCS, a large and growing number of scholars have begun to discuss the value of using place-based and community-engaged approaches under the banner of rhetorical fieldwork.[78] Alongside this work, field histories combine rhetorical field methods and other place-based rhetorical practices with historiography to deeply engage with the ways that (inter)disciplinary genealogies shape "the field(s)" of RWCS. Through field histories, scholars can acknowledge conceptual inheritances, confront relational practices, and overcome spatiotemporal problems. Field histories locate rhetoric within a more dynamic sense of spatial and temporal relationality, as well as engagement within local ecologies of practice.[79] In the archival research and fieldwork I undertook to develop this book, I am less interested in solely presenting the history of ecosystems or ecological inquiry more broadly, which has already been done quite well by Taylor,[80] Phillips,[81] and many others.[82][83] Rather, I want to trace that history in order to understand its influence on the field(s) of RWCS and apprehend how some of ecosystems ecology's knowledge problems became our own, as well as ways in which rhetoric provides some meaningful solutions to those problems.

Far from iconoclasm for its own sake, I believe that field histories lead us to be better ecologists. As Caroline Gottschalk Druschke writes in her definition of trophic ecologies, "The task of the rhetorical ecologist [...] becomes that of co-laboring or equivocating across species, worlds, and registers to take seriously the physicality of relationality."[84] Field histories stand to deepen this sense of co-laboring to better include and emphasize the agential inheritance of our disciplinary past(s). Field histories are intellectual counterhistories that use historiographic methods to place moments of disciplinary change within the context of their paradigms in order to understand how place-times rhetorically shape and influence our contemporary practices and lived experiences within the field. That is, field histories use place-based methods to study synchronic moments of discipline formation and transformation within, but especially against, the narratives through which a discipline makes sense of its fragmented, and often contradictory, past. This method draws on Linda Tuhiwai Smith's notion of "counter stories," which she demonstrates to be "powerful forms of resistance which are repeated and shared across diverse indigenous communities."[85] Field histories also deliberately echo Natasha N. Jones, Kristen R. Moore, and Rebecca Walton's use of antenarrative, which "allows the work of the field [of technical, professional, and scientific communication] to be reseen, forges new paths forward, and emboldens the field's objectives to unabashedly embrace social justice and inclusivity as part of its core narrative."[86] Along similar lines, Aja Y. Martinez's recent book *Counterstory: The Rhetoric and Writing of Critical Race Theory* builds from work in critical race theory to create a method for scholars to undercut master narratives that have abstracted, excluded, and invalidated the perspectives and experiences of minoritized peoples.[87] Counterstory, as Martinez demonstrates, "...effectively turns a reflective mirror on the academy's inherently and institutionally racist histories and environments, which have marginalized and continue marginalizing people of color."[88] Field histories deepen our understanding of rhetoric, providing a richer sense of the events that shape our contemporary practices.[89] Following Skinnell's argument that "contemporary revisionary histories are often pitched toward reinforcing the field's beliefs instead of critically examining them," field histories use a place-based perspective to help us better understand the disciplinary processes of marginalization within both their historical context and their contemporary practice.[90] Using historical field research to understand rhetoric's disciplinary relations may seem, as Druschke claims, "[A]n overwhelming move [but] an ethical one."[91] However, it offers an important means to situate our knowledge practices within deep time and place.

As I develop throughout, ecology as a disciplinary framework, and especially the ecosystem, holds deep-seated conceptual and material connections to the Anthropocene violence of nuclear technology.[92] Contemporary ecology arose in the 1940s through the 1960s from what environmental historian Laura J. Martin refers to as a "history of nuclear colonialism and environmental destruction."[93] In the wake of WWII and the advent of the Cold War, nuclear technology gave rise to the field of radioecology, which produced a fundamental shift in ecology. Patrick Kangas argues that ecological studies of radiation, funded by the US Atomic Energy Commission (AEC), produced a new "conception of the ecosystem" that "allowed cycles of radioactive fallout to be visualized and for radiation effects to be understood," bringing the role of ecological data to the fore.[94] Thus, the move to a quantitative approach is deeply connected to the rise of radioecology. As Joel Hagen puts it, "…[o]ne might also think of a form of symbiosis developing between atomic energy and ecosystems ecology—a relationship in which both partners benefited.[95] The twin concepts of ecology and the Anthropocene are like threads of a rhetorical knot. Together, they function as what Kenneth Burke called "god" and "devil" terms—master-concepts that abstractly structure our ethics.[96] In the following, I draw from decolonial historiography and place-based methods to examine ecology in relation to the Anthropocene not only to acknowledge its disciplinary history of colonial violence but also to work through the spatiotemporal imbroglio that ecology presents as a framework for rhetorical inquiry in a time of global environmental "crisis."

Before embarking on this study of ecosystems and their conceptual afterlives, a few clarifications are needed: in presenting the colonial history of ecology, it may appear that I am condemning science, or ecology specifically, or ecosystems, or rhetoric, or even rhetorical ecologies. While what follows is certainly no hagiography of ecology, this book is not an attempt to level a polemic at science or rhetoric; nor do I aim to criticize, subvert, or deny the laudable and crucial aims of these disciplines. Nor is my intention to denigrate the life and work of notable scientists like the Odum brothers, who made important, meaningful, and lasting contributions to contemporary science and even environmental justice. Rather, I hope that by writing this book I will encourage more people, especially in RWCS, to return to the Odums' work and that of their contemporaries, but this time to read with the lenses provided by rhetoric, historiography, and critical theory. Eugene and H. T. Odum's father Howard Washington Odum was a distinguished scholar who wrote about and advocated for social justice, regionalism, and racial equality. He led the group known as the "Chapel Hill Sociologists" who advocated for modernization and social reform in the American South, and who directly opposed the racist and segregationist views espoused by the Southern Agrarian movement.[97] He instilled in his children from a young age that science was meant to benefit others, rather than to be conducted merely for the sake of advancing knowledge. Further, the Odums are generally beloved in much of the scientific community.

In the time I spent conducting and analyzing interviews with scientists who collaborated with or knew the Odums, and studying their correspondence along with other archival materials, as well as through reading hundreds of papers and reports they and their students wrote, it is quite clear to me that their work has positively impacted the lives of innumerable people for the better and has helped to catalyze ecology into the more vibrant discipline it is today. To be clear, I believe that RWCS scholarship has as much to benefit from studying theories, methods, and practices by scholars like the Odums—and ecological science more generally—as they do in advancing critique. For example, Ariel E. Lugo has written about how Odum used co-authoring practices to support young scientists and to encourage collaboration.[98] Lugo, who began his career in ecology as a student at the University of Puerto Rico, attributes much of his success to Odum's mentorship. Today, Lugo is a prominent ecologist—having directed the International Institute of Tropical Forestry within the United States Department of Agriculture and the Forest Service, served on the board of the Ecological Society of America (ESA), authored more than 470 papers, and received innumerable honors, among many other impressive accolades. In a conversation on Mark Brown's podcast, *The Legacy of H.T. Odum and Systems Ecology*, Lugo reflects on Odum's life and legacy, emphasizing his tremendous contribution as not only a scientist, but also as a teacher, mentor, and collaborator.[99] In that discussion, he lays bare the tensions between the important contributions that H. T. Odum's work in Puerto Rico and the Marshall Islands made and their complicity in the military testing, and its effects, that enabled this work. At the same time, the legacy of Odum's contributions, as represented in his contributions to the El Verde Research Station, persists in shaping disturbance ecology, microbial ecology, invasion ecology, and the carbon cycle, as well as in empowering future scientists at the University of Puerto Rico and the local community.[100] These tensions are always present in Odum's work, and they present scholars in RWCS with important places to acknowledge and understand these influences and agonisms within our own work as scholars, teachers, and practitioners.

While many of the ideas explored during this period of ecology's development did not stand the test of time, technomorphic invention made major contributions to the development of ecology. Ecosystems have helped scientists to conceptualize, identify, and understand the wide range of intractable and interwoven environmental issues we face. I first set out to conduct this research with the firm belief that ecology offers us an essential means of understanding and addressing climate change, and ultimately saving ourselves from annihilation, and I continue to affirm this belief today. Further, the Odums' influence on modern ecology cannot be understated. As Gordon H. Orians puts it in his 1973 book review in *Science*, "Until recently, the appropriate unit of measure of ecology texts was the *odum*, and the problem of selection of a text for a course was a simple one."[101] Because of their iconic influence on ecosystems ecology that carries through to contemporary ecology, I treat their work as a touchstone for understanding how the knowledge problems of their day persist in shaping ecological discourse today. Furthermore, Kuhn and Bruno Latour both spent respective decades fighting against the ways their work in science and technology studies (STS) was mobilized by anti-science interests. I have no desire to revive the so-called science wars. It is not my intention to reproduce the slash-and-burn approach that many scholars of science studies, and their critics, have taken in the past. Nor do I wish to serve as an apologist for the harmful practices that scientists engaged in during this period of time. Rather, this book follows Ryan Skinnell's approach to historiography, which holds that "critique is not a rejection of historical research, nor a rejection of methodological assumptions–it is rigorous examination of the values historians advance, knowingly or not."[102] Historiography, in his model, is "fundamentally iconoclastic."[103] This iconoclastic critique is more a controlled burn than a scorched-earth approach, necessary to promote new growth. By examining ecosystems ecology's ties to nuclear colonialism, I seek to understand the spatiotemporal problems of rhetorical ecologies, and to uncover potential solutions by situating disciplinary practices within lived experiences of time and place.

As a scholar with deep roots in the concept of rhetorical ecologies, I am less interested in leveling criticism at others than I am in doing better myself, to answer Jennifer Clary-Lemon's call for rhetorical scholarship combining decolonial politics and new materialism, asking for "projects that engage differing temporalities (the gifts of past to present, present to future), or terrible inheritances, such as those left in the wake of the Anthropocene."[104] Likewise, I follow Lydia Wilkes' argument that "avowing [...] settler status" makes it possible for "scholars [to] participate in remaking social worlds" toward better and more just futures.[105] As a white, settler-descended scholar, I am directly implicated in scholarly dynamics that credit settler academics with dispossessed peoples' thinking and insights. For example, in "Indigeneity, Posthumanism and Nomad Thought Transforming Colonial Ecologies," Simone Bignall and Daryle Rigney argue that "[c]ontinental posthumanism appears to ignore the prior existence of Indigenous knowledge of this kind,"[106] which threatens "the elision of Indigenous cultural and intellectual authority."[107] Yet the recent work of scholars like Clary-Lemon, who urge new materialist rhetoricians to "examine the move to disassociate from cultural rhetorics and Indigenous knowledges," has convinced me that such a project is needed.[108] In undertaking this work, I follow the example of rhetoricians such as Kristin Arola,[109] David Grant,[110] [111] and Donnie Johnson Sackey et al.,[112] who seek to decolonize the work of "new" materialist rhetorics. As Clary-Lemon argues, "…it is critical for rhetoric and compositionists to imagine the work of new (Indigenous) materialism as equally concerned with decolonization, not only to engage in more ethical and just scholarship, but to better address the knowledge problems of our time."[113] With these caveats and goals in mind, I now turn to rhetoric's ecological moment.

Rhetoric's Ecological Moment

> [W]hen we note that one thinker uses 'God' as his term for the ultimate ground or scene of human action, another uses "nature," a third uses 'environment,' or 'history,' or 'means of production,' etc. And whereas a statement about the grammatical principles of motivation might lay claim to a universal validity, or complete certainty, the choice of any one philosophic idiom embodying these principles is much more open to question.— Kenneth Burke[114]

As what Burke refers to as a "god term," ecology has long influenced what today has become a capacious body of work in RWCS. Burke famously suggested in his *Attitudes Toward History* that "there is one little fellow named Ecology, and in time we shall pay him more attention."[115] In the years that followed, Burke proved to be ahead of his time, being heralded as a pioneer of ecocriticism[116] and a thinker who was influenced by ecosystems ecology's connections with holism and determinism.[117] Alongside Burke, many other rhetoric and composition scholars began to turn to ecology to understand theories, methods, and practices within the field. Starting with Richard Coe's article "Rhetoric 2001,"[118] and then followed the next year by his articles "Eco-Logic for the Composition Classroom"[119] and "Closed System Composition,"[120] ecology entered the purview of rhetoric and writing studies scholars. Around this time, Robert L. Scott also introduced the term "rhetorical environment" to suggest that rhetoric mediates our experiences with places and surroundings.[121] However, it would be a decade later before ecology began gaining real traction as a keyword, marked by the 1986 publication of Marilyn Cooper's influential article "The Ecology of Writing,"[122] followed the next year by Karen Burke LeFevre's *Invention as a Social Act*.[123] These works brought about the emergence of what Byron Hawk refers to as a "counter-tradition" in RWCS, "…overshadowed by the social turn as the field's dominant paradigm during the 1980s and 1990s."[124] From there, ecology began to solidify as a disciplinary focus in rhetoric with works like Louise Wetherbee Phelps' book *Composition as a Human Science*,[125] and the concept became more central to the field with Margaret Syverson's book *The Wealth of Reality*.[126] Jenny Edbauer's work with public discourse and rhetorical ecologies further cemented ecology as a rhetorical framework.[127]

In the early 2000s, as Hawk explains, ecology moved to the "forefront of rhetoric and composition,"[128] initiating what many scholars now characterize as an "ecological turn" in RWCS[129][130][131] and has even been referred to recently as a "subdiscipline" of rhetoric.[132] In his previous work, Hawk traces counterhistories of composition, including the relationship between networks and ecology.[133] In *Rhetorical Ecologies*, we present a broad conceptual history of rhetorical ecologies as it developed across the field(s) of writing studies (including rhetoric and composition studies) broadly.[134] Specifically we trace its origins with Coe through its various and wide-ranging connections to ecocomposition,[135] pedagogy,[136] hypermedia,[137] new media interfaces,[138] visual rhetoric,[139] geographical perspectives on writing,[140] genre,[141][142] composition,[143] writing (post)process,[144] assessment,[145] writing program administration,[146] circulation,[147] writing studies research,[148] and digital rhetoric,[149][150] to name but some of the numerous writing studies scholars who have evoked ecology. In their twenty-year retrospective analysis of a writing studies journal, Noah Patton and Rachel Presley "uncovered the critical function that rhetorical ecologies play in *Reflections*' production, circulation, and sustained value."[151] In the introduction to their edited collection *Technological Ecologies and Sustainability*, Dickie Selfe, Dànielle Nicole DeVoss, and Heidi A. McKee introduce "technological ecologies" to describe "digital environments" and networks in ways that directly evoke the technomorphic tensions present in ecosystems ecology.[152] As the wide range of contributions in the collection make clear, EH and DH methods and topics are deeply connected in writing studies scholarship through the connections between ecology and technology.

Ecology has also been important to research in rhetorical scholarship that examines systems theory in relation to writing,[153] [154] the trophic dynamics of rhetorical ecologies,[155] field methods,[156] the rhetoric of health,[157] the ecologies of data centers,[158] the rhetoric of environmental guilt,[159] and even to define rhetorical ecologies as ontologically distinct from the scientific discipline.[160] Other scholars in writing studies have critiqued rhetorical ecological metaphors from the perspective of ecological literacy[161] and called for a turn toward land-based ecological inquiry.[162] Angela Haas has used decolonial methodologies to critique rhetorical ecologies in relation to DH as part of "theorizing a social justice approach to ecological literacies."[163] Haas identifies many of the issues discussed in this book, from exploration and exploitation to frontierism and pioneerism, and she critiques the "digital nativism" metaphor.[164] Along similar lines, Lou Maraj theorizes an antiracist "deep rhetorical ecology" or "deep ecology" to describe an "evolving series of rhetorical situations in which communication occurs, which are interrelated through bodies, spaces, cultures, and contexts with specific regard to power dynamics and race relations."[165] In his work with data centers and critical infrastructure literacies, Dustin Edwards criticizes the view of big data as "simply digital information circulating in some abstracted techno-ecology" as obscuring its material environmental impacts.[166] In addition to the breadth of writing studies scholarship focusing on rhetorical ecologies, a large number of scholars in environmental communication have written about rhetorical ecologies and adjacent concepts.[167]

All of this to say: for RWCS, ecology has increasingly become a threshold concept in contemporary scholarship, offering a rhetorical framework that indexes the study of networked discourse, new materialism, systems thinking, and many other important areas of focus. In 2012, Noah Roderick proposed that, for writing studies, "…complexity and ecology are rapidly becoming dominant metaphors."[168] Today, scholarship on rhetorical ecologies, ecocomposition, and environmental communication has become as capacious and expansive as the concept itself, and ecology has grown into what we might characterize as a recognizable meta-discipline or subfield of RWCS and a cornerstone of EH/DH. However, as Dan Ehrenfeld notes in his recent critique of rhetorical ecologies, "[E]cological models have emphasized flux" but in doing so "they have deemphasized historical specificity."[169] While Ehrenfeld turns away from ecology as a framework and toward infrastructural models, this book follows Donna Haraway's notion of "staying with the trouble" of ecology and offers a method for rhetorical inquiry to further emphasize the historical elements of any framework, not only as a corrective to the ahistorical problems Ehrenfeld identifies, but also to challenge the colonial ideology and disciplinary narratives on which rhetorical ecologies rest.[170] Studying the colonial history of ecology alongside atomic and Anthropocene temporalities helps to situate ecology's spatiotemporal concerns within the violent displacements and derangements of Eurocentric spatial and temporal scales.[171]

Whether referring to energy moving through a biological community or information circulating in a digital network, the term "ecology" now connotes many different types of relational systems. These systems cross the biological, technological, and ideological with the virtual and material in what Guattari terms the "three ecologies."[172] To invoke ecology is to gesture toward a host of ambiguous associations: complexity, scale, dynamics, boundaries, systems, emergence, and flux to name but a few. While ecology as we know it today is regarded as a master discipline, structuring the way we see the world, its emergence in public discourse is a distinctly modern development. Ecosystems ecology first arose in the 1950s and 1960s from the nuclear tests initiated in the 1940s and went on to reshape ecology as a discipline. Leading up to this, ecology began to solidify as a "hard" scientific discipline when Eugene Odum and his brother Howard borrowed structuring metaphors from economics and cybernetics to theorize a systems approach to the emerging field.

While Tansley and Lindeman laid the foundations of the ecosystem, this approach was significantly developed by G. Evelyn Hutchinson who was Howard T. Odum's dissertation director. For his pioneering work, Eugene Odum is lionized as "the father of modern ecology." The Odum School of Ecology at the University of Georgia is named for Eugene (who founded the program and taught there from 1940 to 1984), as is the ESA's Eugene P. Odum Award for Excellence in Ecology Education. Howard T. Odum is likewise celebrated as one of the founders of modern ecology, as well as a pioneer in ecological engineering, modeling, and economics. The Howard T. Odum Center for Wetlands at the University of Florida is likewise named for its founder, who worked at UF from 1970 until 1996. The Odum Award, named in recognition of both Odum brothers, is the highest honor bestowed by the American Ecological Engineering Society. The Odum brothers' work with the mesocosm and closed systems was a major influence on the design of Biosphere 2, a subject I return to in Chap. 4. Together, the Odum brothers were awarded the prestigious Crafoord Prize (akin to a Nobel Prize in bioscience) from the Royal Swedish Academy of Sciences in 1987.

At the end of WWII, with generous support from the AEC, this state-of-the-art ecological research was fueled, in large part, by the interests of nuclear colonialism.[173] Thus, the paradigm shift toward contemporary ecology, what Carolyn R. Miller terms a kairotic "opening" of intellectual terrain,[174] is directly linked to the more than 65 nuclear bombs exploded in the Marshall Islands, where the Odum brothers conducted their groundbreaking fieldwork (Fig. 1.1).

This rhetorical moment, this "ecological turn," pivoted on the nuclear colonial violence perpetrated against the Marshall Islanders by the US Army. While fears of nuclear annihilation spurred US funding for ecological studies like those the Odums conducted at the Enewetak Atoll, the Marshall Islanders lived through actual nuclear violence, suffering "forced relocations, destruction of ancestral lands, and radiation sickness."[175] By engaging with this history, this book places the "spectacular violence and mundane resistance" of what Megan Eatman terms "violent rhetorical

Fig. 1.1 Aerial image of the July 1, 1946, Able detonation at the Bikini Atoll (United States Army Air Forces), the first detonation of Operation Crossroads (consisting of Able, Baker, and Charlie). Retrieved from Wikimedia Commons, https://commons.wikimedia.org/wiki/File:Operation_Crossroads_-_Able_001.jpg (Public Domain)

ecologies"[176] within the nuclear colonial event of the Anthropocene as it continues to unfold at the Pacific Proving Grounds. Operation Crossroads resulted in both material and theoretical violence against the Marshallese people and their Lands that continues to this day. In the Castle Bravo detonation, which turned out to be hundreds of times more powerful than estimated, the people of Enewetak were exposed to radioactive fallout and stranded for days on the islands where they had been relocated. When the US military selected Bikini Atoll as a site for their tests, the people of Bikini were forced to relocate to the desolate Rongerik Atoll, where resource scarcity brought starvation and dehydration. After two years, they were again relocated to Kwajalein and then to Kili Atoll. As historians of science have pointed out, these relocations created numerous immediate hardships for the people of Bikini, but the loss of their homeland also caused "the loss of skills required for self sustenance."[177] The immediate violence of displacement resulted in ongoing cultural violence as the islanders found that their place-based fishing practices would no longer adequately support their life on Kili, in addition to the ongoing material effects of radioactive exposure and the obliteration of their traditional homelands.

As Martin demonstrates, a theoretical violence also emerged from this material history as it became entwined with the Odums' ecological fieldwork at Enewetak. As I show in the following, these events helped shape ecological science by providing a place to study ecosystems through extractive field-research practices. This violence is directly connected to the exploitative "frontier of science" metaphor that Leah Ceccarelli critiques in contemporary American science.[178] Likewise, the reliance of ecosystems ecology on nuclear pollution mirrors the critiques that Max Liboiron (Métis/Michif) forwards in *Pollution Is Colonialism*, where she argues that pollution is an enactment of persisting colonial Land relations.[179] She argues that pollution itself, which in Liboiron's study refers to plastic, makes colonial science possible, and that this science is in turn used to justify "acceptable" levels of pollution. Clearly, her arguments map directly onto nuclear radiation pollution and the relationship between energy and radioecology. Just as nuclear pollution has been viewed as a marker that makes the Anthropocene visible as a (now rejected) geological period, so did nuclear radiation serve to make ecological systems visible in the Odums' research projects. The afterlife of these studies continues to impact the Lands and peoples today. Danielle Endres' *Nuclear Decolonization: Indigenous Resistance to High-Level Nuclear Waste Siting* documents Indigenous activism that resists nuclear colonization. Endres shows that "Indigenous peoples and nations disproportionately bear the burdens and harmful effects of the nuclear production process, in most cases imposed by settlers."[180] Liboiron and Endres both use the term *Lands* to refer to "the entire ecology of beings present in Indigenous territories."[181] As Enewetak participates in the "discovery" or "invention" of the ecosystem, the conditions of nuclear colonialism, its impact on the Indigenous Lands, peoples, and nations, as well as overt military funding become part of the ecological paradigm shift. In turn, I argue that rhetorical ecologies have inherited this conceptual history of Anthropocene violence from the influence of nuclear colonialism on ecosystems ecology. My aims in recognizing and reckoning with this violent history of ecology are twofold: (1) to acknowledge and emphasize the ways that the ecological framework has marginalized both the spectacle of nuclear violence and the stories and lived experiences of communities that were subjected to—and continue to resist—the violence of nuclear colonialism, and (2) to come to terms with the theoretical, material, and practical constraints that this spatiotemporal inheritance places on contemporary rhetoric and ecological inquiry.

Building from this exigency, the last chapter of the book discusses the community-based work mobilized by the Digital Writing Environments, Location, and Localization (DWELL) Lab at the University of Rhode Island that I founded in 2020. Our focus on community-based and participatory science communication projects is inspired by the cutting-edge work that Liboiron and collaborators are doing in the Civic Laboratory for Environmental Action Research (CLEAR), an "anticolonial lab" which focuses on plastic pollution by studying the guts of marine animals.[182] Liboiron explains that the anticolonial practices of the lab "place land relations at the centre of our knowledge production as we monitor plastic pollution in the province of Newfoundland and Labrador."[183] For example, CLEAR conducts its research without toxic chemicals[184] or those that "require hazardous waste disposal," as the use of such chemicals "assumes access to Land as a sink."[185] They also use "judgmental sampling rather than random sampling" in order to "foreground food sovereignty when we look at plastics in food webs."[186] These practices constrain the kinds of research that CLEAR can do, but they also foreground the ways that colonialism often shapes scientific practices and show how researchers can learn to resist those assumptions. Following their work, the final chapter will discuss examples of the community-based projects that DWELL collaborators have been producing in recent years.

Progression of the Book

The modern form of the ecosystem concept was solidified in the work of Eugene and Howard T. Odum. Following the historical and theoretical discussions of ecosystems covered in this introduction, Chaps. 2 and 3 take a closer look at the ecosystem as a diagram and apparatus, examining the role that the rhetoric of energy and cybernetics (information theory), specifically regarding nuclear radiation, and large-scale radioecology experiments played in making the ecosystem possible. Chapter 2 specifically focuses on the rhetoric of energy in H. T. Odum's famous study at Silver Springs, which first mapped trophic dynamics in a freshwater ecosystem in central Florida. Bridging from George Kennedy's famous definition of rhetoric as a form of energy exchange, this chapter examines the relationship between energy and rhetoric as they circulate through networks and places. Chapter 3 investigates radioecology at the proving grounds of the ecosystem, namely at the Enewetak Atoll research site in the Marshall Islands and at the El Verde Field Station in Río Grande, Puerto Rico. This chapter extends the discussion of energy rhetoric to examine how nuclear radiation served to make the ecosystem concept visible while negatively impacting Indigenous Lands, peoples, and nations.

Chapters 4 and 5 serve as case studies for the theoretical and methodological concerns introduced in previous chapters. Chapter 4 examines how problems of spatial scale, intractable issues that are deeply rooted in the colonial history of ecosystems, affect contemporary environmental inquiry. This chapter uses the rhetorical concept of *megethos* (or "magnitude" and "greatness") to understand how scale limits and delineates both rhetoric and ecological inquiry. Building from this work, Chap. 5 draws upon the rhetorical concept of *kairos* (alongside "deep time") to discuss how problems of time and temporality impact environmental inquiry. Together, these chapters illustrate the practical implications of the history discussed in previous chapters, offering careful analysis based on archival research and fieldwork in places where the Odums conducted their studies. In Chap. 6, I conclude by reflecting on how the intellectual history I outline throughout the book shapes knowledge problems that environmental communicators face today. I demonstrate how RWCS scholars, teachers, and practitioners can use the theories and methods I have discussed to inform the work they do. I draw examples from the community-based work being done in the DWELL Lab at the University of Rhode Island to demonstrate how we can foster more sustainable, dialogical, and responsive approaches to environmental communication. Taken together, this is a book about staying with the trouble of ecology by tracing its conceptual history, examining its material impacts, and working together with communities to resist the persistent issues this history has given us in order to build toward better futures. Ecosystems bridge between concepts of media, environments, and information networks; energy and pollution; places, peoples, and times. As I discuss throughout this book, ecosystems hold the potential to make systems of oppression visible and tangible, and in doing so, they compel us toward alterity.

Notes

1. E. M. Forster, "The Machine Stops," *The Oxford and Cambridge Review* 8 (1909): 96.
2. Ralph Jessop, "Coinage of the Term Environment: A Word without Authority and Carlyle's Displacement of the Mechanical Metaphor," *Literary Compass* 9, no. 11 (2012): 708–720, doi:10.1111/j.1741-4113.2012.00922.x
3. Ibid., 710. As Jessop demonstrates, the coinage of the term environment occurs "within a broader narrative of the transmission of organicist, antimechanical, counter-Enlightenment discourses, bringing the notion of environment into relation with a much more extensive story of later attempts to undermine the authority or prevalence of mechanism by writers, thinkers, composers, artists, and campaigners throughout the nineteenth and twentieth centuries."
4. Ernst Haeckel, *Generelle Morphologie der Organismen: Allgemeine Grundzüge der Organischen Formen; Wissenschaft, Mechanisch Begründet durch die von Charles Darwin Reformierte Deszendenz-Theorie* (G. Reimer, 1866).
5. Thomas Kuhn, *The Structure of Scientific Revolutions* (The University of Chicago Press, 1962), 66.
6. Arthur G. Tansley, "The Use and Abuse of Vegetational Concepts and Terms," *Ecology* 16 (1935): 284–307, doi:https://doi.org/10.2307/1930070
7. Thomas Pringle, Gertrud Koch, and Bernard Stiegler, *Machine* (Meson Press, 2019), 62. As Pringle, Koch, and Stiegler explain in *Machine*, Tansley's ecosystem was partially inspired by Sigmund Freud's notion that the human brain was made up of an electrical network. Tansley underwent psychoanalytic treatment twelve years before he published his study, and throughout that time he "remained fascinated with Freud's

theory and practice throughout his career, even laying out his own theory of mind in resemblance of his concept of the ecosystem."
8. Peter Ayres, *Shaping Ecology: The Life of Arthur Tansley* (Wiley-Blackwell, 2012), 138; A.J. Willis, "Arthur Roy Clapham, 24 May 1904–18 December 1990," *Biographical Memoirs of Fellows of the Royal Society* 39 (1994): 71–90, doi:10.1098/rsbm.1994.0005
9. See ibid., 138 for more information on the relationship between Clapham and Tansley as well as Clapham's biographical entry from The Royal Society, A. J. Willis, "Arthur Roy Clapham, 24 May 1904–December 1990," *Biographical Memoirs of Fellows of the Royal Society* 39, (1994): 71–90, doi:10.1098/rsbm.1994.0005
10. Peder Anker, *Imperial Ecology Environmental Order in the British Empire, 1895–1945* (Harvard University Press, 2001), 7. He explains that by using "Freudian theories of the human mind and social psychology" as a point of departure, Tansley "explored how the aims and methods of ecological research could enhance British imperialism."
11. Benjamin Golley, *A History of the Ecosystem Concept in Ecology* (Yale University Press, 1993), 2.
12. Gottfried Schnödl and Florian Sprenger, *Uexküll's Surroundings: Umwelt Theory and Right-Wing Thought* (Meson Press, 2021), 12.
13. Thomas Pringle, "The Tech Ecosystem and the Colony," *Heliotrope Journal*, May 12, 2021, par. 2, https://www.heliotropejournal.net/helio/the-tech-ecosystem. Anker argues that this agonism led to a "clash between Tansley and Smuts's protégé John Phillips" that "was perhaps the most important ecological debate of the period." Anker, *Imperial Ecology*, 2.
14. Plato, Sophist (245d); qtd. in J.C. Smuts, *Holism and evolution* (Macmillan, 1926), v.
15. Ibid. Surprisingly, before Smuts drafted the 1910 manuscript, *An Inquiry into the Whole*—which was published in 1926 under the title *Holism and Evolution*—he wrote a book-length study titled *Walt Whitman: A Study in the Evolution of Personality* that remained unpublished until 1973, years after his death. Such a fact suggests the deep-seated ways that poetics and linguistics influenced how Smuts imagined the relationship between his views on holism, evolution, and white supremacy.
16. Greg Garrard, "Heidegger Nazism Ecocriticism," *ISLE: Interdisciplinary Studies in Literature and Environment* 17, no. 2 (2010): 251–271, doi:10.1093/isle/isq029; Enrich Hörl, *General Ecology: The New Ecological Paradigm* (Bloomsbury, 2017), n19, 48. For more information on the complex historical relationship between environmental theory and fascism, see Greg Garrard's "Heidegger Nazism Ecocriticism." Enrich Hörl distinguishes between Environmentality "with a capital 'E'"

to suggest the work of Michel Foucault and Brian Massumi, while the "small 'e'" suggests "the concept *Umweltlichkeit* as used by Heidegger in *Being and Time*." Such divergence indicates how neither concept is removed from a political and ethical context.
17. Elsewhere, this concept has been referred to as "mechanomorphism" and "robotic anthropomorphism." See Linnda R. Caporael, "Anthropomorphism and Mechanomorphism: Two Faces of the Human Machine," *Computers in Human Behavior* 2, no. 3 (1986): 215–234, doi:10.1016/0747-5632(86)90004-X; Carl DiSalvo, Francine Gemperle, Jodi Forlizzi, and Sara Keisler, "All Robots Are Not Created Equal: The Design and Perception of Humanoid Robot Heads," in *Proceedings of the 4th conference on Designing interactive systems: processes, practices, methods, and techniques (DIS '02)*, (2002), doi:10.1145/778712.778756; Carl DiSalvo and Francine Gemperle, "From Seduction To Fulfillment: The Use of Anthropomorphic Dorm in Design," in *Proceedings of the 2003 International Conference on Designing Pleasurable Products and Interfaces (DPPI '03)*, (2003), doi:10.1145/782896.782913
18. Janet Vertesi, *Seeing Like A Rover: How Robots, Teams, and Images Craft Knowledge of Mars* (The University of Chicago Press, 2015). Vertesi examines how technomorphism shapes the work being done by a team of engineers assigned to the *Spirit* and *Opportunity* Mars Rover missions.
19. Tsvi Tlusty and Albert Libchaber, "Life Sets Off a Cascade of Machines," *Proceedings of the National Academy of Sciences* 122, no. 4 (2025): e2418000122, doi:10.1073/pnas.2418000122. Tlusty and Libchaber posit that life self-organizes, or "cascades," into machine-like systems, scaling up from the atomic level to the level of the biosphere.
20. Jim Thatcher and Craig M. Dalton, *Data Power: Radical Geographies of Control and Resistance* (Pluto Press, 2021).
21. Nicholas Van Horn, Aaron Beveridge, and Sean Morey, "Attention Ecology: Trend Circulation and the Virality Threshold," *Digital Humanities Quarterly* 10, no. 4 (2016), https://www.digitalhumanities.org/dhq/vol/10/4/000271/000271.html Today, we see technomorphism in examples like emissions budgets and carbon credits, the food chain, references to burnout, feeling wired, or "hacking" the human body, as well as metaphors in biological networks, such as the "Wood Wide Web" or mycorrhizal networks. While examples like "sharp witted" (as in keen) or "honing" an argument have etymological roots in references to weapons, or the phrases "cog in the machine" and "like clockwork" compare humans to machines, other examples like "machine learning," "political systems," and "artificial intelligence," as well as the use of virality as a metaphor for human attention online, further tangle

the threads of anthropomorphism and technomorphism. S. Scott Graham, *The Doctor and the Algorithm: Promise, Peril, and the Future of Health AI* (Oxford University Press, 2022). Graham's *The Doctor and the Algorithm* offers an example of how AI in health discourse often reproduces systemic inequities based on human bias and which complicate distinctions between anthropomorphism and technomorphism.

22. See John Lynch and William J. Kinsella, "The Rhetoric of Technology as a Rhetorical Technology," *Poroi* 9, no. 1 (2013): 13, 1–6, doi: https://doi.org/10.13008/2151-2957.1152

23. Lawrence Prelli, *A Rhetoric of Science: Inventing Scientific Discourse* (University of South Carolina, 1989).

24. For example, rhetoricians of science have long studied the role of genre in science communication. See A. G. Gross, *The Rhetoric of Science* (Harvard University Press, 1990). RSTM has long studied the role of genre in science communication. See Alan G. Gross, Joseph E. Harmon, and Michael Reidy, *Communicating Science: The Scientific Article from the seventeenth Century to the Present* (Oxford University Press, 2002); Charles Bazerman. 1989. "What Are We Doing as a Research Community? Introduction." *Rhetoric Review* 7 (2): 223–224, doi:10.1080/07350198909388856 RSTM scholars have also focused on issues of boundary work and demarcation within scientific fields. See Thomas F. Gieryn, "Boundary-Work and the Demarcation of Science from Non-Science: Strains and Interests in Professional Ideologies of Scientists," *American Sociological Review* 48, no. 6 (1983): 781–795, doi:10.2307/2095325; Charles Alan Taylor, *Defining Science: A Rhetoric of Demarcation* (University of Wisconsin Press, 1996). Likewise, RSTM scholars have focused on the rhetorical construction of scientific controversy and authority in science. See S. Scott Graham and Lynda Walsh, "There's No Such Thing as a Scientific Controversy," *Technical Communication Quarterly* 28, no. 3 (2019): 192–206, doi:10.1080/10572252.2019.1571243; A. G. Gross, "On the Shoulders of Giants: Seventeenth-century Optics as an Argument Field," in *Landmark Essays on Rhetoric of Science: Case Studies*, ed. R. A. Harris (Lawrence Erlbaum Associates, Inc., 1997), 19–38; M. M. Marquardt, "Creationist Science and the Rhetorical Capacity of the Scientific Method," *Rhetoric Review* 41, no. 2 (2022): 130–145, doi:10.1080/07350198.2022.2038508; Lynda Walsh, *Scientists as Prophets: A Rhetorical Genealogy* (Oxford University Press, 2013).

25. For a history of the development of rhetorical new materialism (RNM), see S. Scott Graham, *Where's the Rhetoric? Imagining a Unified Field* (Ohio State University Press, 2020). Graham discusses important recent examples of RNM, including: Thomas Rickert, *Ambient Rhetoric: The*

Attunements of Rhetorical Being (University of Pittsburgh Press, 2013); L. E. Gries, *Still Life With Rhetoric: A New Materialist Approach for Visual Rhetorics* (Utah State University Press, 2015); Christa Teston, *Bodies in Flux: Scientific Methods for Negotiating Medical Uncertainty* (University of Chicago Press, 2017); S. Barnett and C. Boyle, *Rhetoric, Through Everyday Things* (University of Alabama Press, 2016); Marilyn Cooper, *The Animal Who Writes: A Posthumanist Composition* (University of Pittsburgh Press, 2019); S. Scott Graham, *The Politics of Pain Medicine: A Rhetorical-Ontological Inquiry* (University of Chicago Press, 2015); J. A. Lynch and N. Rivers, *Thinking with Bruno Latour in Rhetoric and Composition* (Southern Illinois Press, 2015); Casey Boyle, *Rhetoric as a Posthuman Practice* (Ohio State University Press, 2018). Casey Boyle points to the field's interest in Bruno Latour's actor-network theory (ANT) and Donna Haraway's "cyborg manifesto" as popular works which contributed to the development of RNM, but he also offers a more expansive history that includes the intellectual contributions of much older works like Bergson's *Matter and Memory* and Burke's *Permanence and Change*. Bruno Latour, *Reassembling the Social: An Introduction to Actor Network Theory* (Oxford University Press, 2005); Donna Haraway, "A Manifesto for Cyborgs: Science, Technology, and Socialist Feminism in the 1980s," *Socialist Review* 80 (1985): 15(2): 65–107; H. Bergson, *Matter and Memory* (Macmillan, 1986); Kenneth Burke, *Permanence and Change: An Anatomy of Purpose* (New Republic, 1935).

26. Sidney I. Dobrin and Christian Weisser, *Ecocomposition: Theoretical and pedagogical approaches* (SUNY Press, 2001). Sidney I. Dobrin and Christian Weisser, *Natural Discourse: Toward Ecocomposition* (SUNY Press, 2002). Sidney I. Dobrin and Christian Weisser, "Breaking Ground in Ecocomposition: Exploring Relationships between Discourse and Environment," *College English* 64, no. 5 (2002): 566–589, doi:10.2307/3250754 For example, see Sidney I. Dobrin and Christian R. Weisser's three-part initiative, beginning with their edited collection *Ecocomposition: Theoretical and Pedagogical Approaches* (SUNY Press, 2001); and followed by *Natural Discourse: Toward Ecocomposition* (SUNY Press 2002); and their article "Breaking Ground in Ecocomposition: Exploring Relationships between Discourse and Environment," *College English* 64, no. 5 (2002): 566–589 https://doi.org/10.2307/3250754

27. For an example of this division, see: Carolyn R. Miller, "*Kairos* in the Rhetoric of Science," in *A Rhetoric of Doing: Essays Honoring James L. Kinneavy*, Steven P. Witte et al., eds. (Southern Illinois University Press, 1992), 310–27; Carolyn R. Miller, "Opportunity, Opportunism, and Progress: *Kairos* in the Rhetoric of Technology," *Augmentation* 8 (1994): 81–96, doi:10.1007/BF00710705

28. Ira Allen, *Panic Now? Tools for Humanizing* (University of Tennessee Press, 2024), 16.
29. Philip Eubanks, *The Troubled Rhetoric and Communication of Climate Change: The Argumentative Situation* (Routledge, 2015). For instance, Eubanks' *The Troubled Rhetoric and Communication of Climate Change* examines climate change discourse through RWCS perspectives.
30. Debra Hawhee, *A Sense of Urgency How the Climate Crisis Is Changing Rhetoric* (University of Chicago Press, 2023); Joshua Trey Barnett, *Mourning in the Anthropocene: Ecological Grief and Earthly Coexistence* (Michigan State University Press, 2022); Jennifer Clary-Lemon, *Nestwork: New Material Rhetorics for Precarious Species* (Penn State University Press, 2023). For example, Hawhee's *A Sense of Urgency* describes the ways that crisis is changing the ways we communicate and persuade, Barnett examines emotional responses to the anthropocene, and Clary-Lemon examines forms of communication between humans and nonhumans, particularly with species that depend on human development for habitat.
31. Ehren Helmut Pflugfelder, *Geoengineering, Persuasion, and the Climate Crisis: A Geological Rhetoric* (University of Alabama Press, 2022).
32. Caroline Gottschalk Druschke and Bridie McGreavy, "Why Rhetoric Matters for Ecology," *Frontiers in Ecology and Environment* 14, no. 1 (2016): 46–52, doi:10.1002/16-0113.1
33. Nathan Stormer et al., *Rhetorical Climatology: By A Reading Group* (Michigan State University Press, 2023).
34. See Kathryn M. Northcut, "Stasis Theory and Paleontology Discourse." *The Writing Instructor*, 3 (2007), http://www.writinginstructor.com/northcut. Northcut applies stasis theory to understand science discourse in a paleontology debate.
35. Erika Amethyst Szymanski, "Constructing Relationships Between Science and Practice in the Written Science Communication of the Washington State Wine Industry," *Written Communication* 33, no. 2 (2016): 184–215, doi: 10.1177/0741088316631528. Erika Amethyst Szymanski, "What Is the Terroir of Synthetic Yeast?" *Environmental Humanities* 10, no. 1 (2018): 40–62, doi:10.1215/22011919-4385462. Erika Szymanski (2016, 2018) has written extensively on the subject of human–microbe relations and science communication through the perspective of rhetoric and STS. See also Ehren Helmut Pflugfelder, "Rhetoric's New Materialism: From Micro-Rhetoric to Microbrew." *Rhetoric Society Quarterly* 45, no. 5 (2015): 441–461. doi: 10.1080/02773945.2015.1082616
36. Anker, *Imperial Ecology*, 1. Anker explains that "[t]he formative formative period of ecological reasoning coincides with the last years of the

British Empire" and as such the discipline directly "grew out of the imperial administrative and political culture."
37. Joel Hagen, *An Entangled Bank: The Origins of Ecosystems Ecology* (Rutgers University Press, 1992), 80.
38. Dana Phillips, The Truth of Ecology: Nature, Culture, and Literature in America (Oxford University Press, 2003).
39. Enrich Hörl, *General Ecology: The New Ecological Paradigm* (Bloomsbury, 2017).
40. Thomas Pringle, "The Tech Ecosystems and the Colony," *Heliotrope Journal*, May 12, 2021, par. 3, https://www.heliotropejournal.net/helio/the-tech-ecosystem
41. Peter J. Taylor, "Technocratic Optimism, H. T. Odum, and the Partial Transformation of Ecological Metaphor after World War II," *Journal of the History of Biology* 21, no. 2 (1988): 213, doi:10.1007/BF00146987
42. Howard Scott, *Science* Versus *Chaos!* (Technocracy Inc., 1933).
43. Barry Jones, *Sleepers, Wake!: Technology and the Future of Work* (Wheatsheaf Books, 1982), 210. Jones' *Sleepers, Wake!* provides a detailed discussion of technological determinism in relation to the technocracy movement.
44. Peter J. Taylor, *Unruly Complexity Ecology, Interpretation, Engagement* (The University of Chicago Press, 2005), 52–54. For an in-depth discussion of the influence of the technocracy movement and its influences on G. Evelyn Hutchinson and his student H. T. Odum (among many others), see Taylor's *Unruly Complexity.*
45. Enrich Hörl, *General Ecology: The New Ecological Paradigm* (Bloomsbury, 2017), 2, 5.
46. E. M. Forster, "The Machine Stops," *The Oxford and Cambridge Review* 8, (1909).
47. Nicholas D. Holland, "Walter Garstang: A Retrospective," *Theory in Biosciences* 130 (2011): 248, doi:10.1007/s12064-011-0130-3; J. Butler Burke, "On Haeckel and Haeckelism," *The Oxford and Cambridge Review* 2 (1907): 18–35. In fact, the essay "Haeckel and Haeckelism" appeared in a 1907 issue of *The Oxford and Cambridge Review*, two years before "The Machine Stops."
48. Edward Henry Howell, "Modernism, Ecology, and the Anthropocene" (PhD diss., Temple University, 2017), 19, Temple University Library, 20.500.12613/1466. Howell proposes the term to refer to a "literary history" that begins "at the tail end of the age of empire" and closes "in the middle of the twentieth-century."
49. Marcin Tereszewski, "Dystopian Space in E. M. Forster's 'The Machine Stops,'" *Language and Literary Studies* 10, (2020): 225–236, https://www.proquest.com/docview/2575546810
50. E. M. Forster, *Howards End* (Edward Arnold, 1910).

51. E. M. Forster, "The Machine Stops," *The Oxford and Cambridge Review* 8 (1909): 83–122. Another interesting connection occurs in *Howards End*, when a character mentions having dinner at Eustace Miles restaurant, which Tansley frequented daily in his time at University College, London. Anker, *Imperial Ecology, 12*; Howell, "Modernism, Ecology," 4. This restaurant was a popular meeting place for "socialists and London's counterculture" Anker, *Imperial Ecology*, 12.
52. Alf Seegert, "Technology and the Fleshly Interface in Forster's 'The Machine Stops': An Ecocritical Appraisal of a One-Hundred Year Old Future," *Journal of Ecocriticism* 2, no.1 (2010): 48, https://ojs.unbc.ca/index.php/joe/article/view/98
53. Dana Phillips, *The Truth of Ecology: Nature, Culture, and Literature in America* (Oxford University Press, 2003); Jean Baudrillard, *Simulations* (Semiotext(e), 1983). Phillips builds from Jean Baudrillard's concept of the "hyperreal"—where experience becomes highly simulated to the point that it becomes at once fake and also more real than reality—to understand some of the conceptual problems facing ecocriticism at the time. Namely, Phillips charges ecocritical thinking with moving beyond the pastoral trappings of modernism and coming to terms with postmodern theory.
54. Casey Boyle and Nathaniel Rivers, "Augmented Publics," in *Circulation, Writing, and Rhetoric*, L. E. Gries and C. G. Brooke, eds. (University Press of Colorado, 2018), 83–101, doi:10.2307/j.ctt21668mb.8. Boyle and Rivers refer to the ways that technologies and places, especially in the relationship between the two, cultivate "a public as a circulatory project" which necessitates rhetorical perspectives to understand our increasingly networked public sphere.
55. Adam Curtis, dir., *All Watched Over by Machines of Loving Grace*. 1, 2, "The Use and Abuse of Vegetational Concepts." Aired May 30, 2011, on BBC.
56. Richard Brautigan, *All Watched Over by Machines of Loving Grace* (Communication Company, 1967), 67.
57. Nichole Starosielski, *The Undersea Network* (Duke University Press, 2015).
58. Neal Stephenson, "Mother Earth Mother Board," *Wired Magazine*, December 1996.
59. Starosielski, *The Undersea Network*, 171.
60. Ibid., 172.
61. Ibid., 172.
62. Ibid., 173.
63. Sidney I. Dobrin and Madison Jones, eds., *Rhetorical Ecologies* (National Council of Teachers of English Press, 2024), 1.
64. Félix Guattari, *The Three Ecologies* (Continuum, 2008), 43.

65. Gilles Deleuze and Félix Guattari, *A Thousand Plateaus: Capitalism and Schizophrenia* (University of Minnesota Press, 1987).
66. Eugene P. Odum, *Ecology; The Link Between the Natural and the Social Sciences* (Holt, Rinehart and Winston, 1975), 122.
67. A few examples are Gregory Bateson, *Steps to an Ecology of Mind* (University of Chicago Press, 1972); Yves Citton, *The Ecology of Attention* (Polity, 2017); Paul Robbins, *Political Ecology: A Critical Introduction* (Blackwell Publishing, 2004); Alicia Carroll, *New Woman Ecologies: From Arts and Crafts to the Great War and Beyond* (University of Virginia Press, 2019); Scott Hess, *William Wordsworth and the Ecology of Authorship* (University of Virginia Press, 2019).
68. John C. Tinnell, "Transversalising the Ecological Turn: Four Components of Félix Guattari's Ecosophical Perspective," *Deleuze Studies* 6, no. 3 (2012): 357–388, http://www.jstor.org/stable/45331514. Margaret Linley, "Ecological Entanglements of DH," in *Debates in the Digital Humanities*, M. K. Gold and L. F. Klein, eds. (University of Minnesota Press, 2016), 411. Diana Keeling, "Of Turning and Tropes," *Review of Communication* 16, no. 4 (2016): 317. Tinnell discusses the "ecological turn" across English studies broadly. Linley discusses "the ecological turn in the digital humanities." Keeling examines "the turn" as a trope that factors largely into "the production of rhetoric's intellectual history."
69. Pringle, "The Tech Ecosystems," para. 3.
70. Smuts, *Holism and evolution*.
71. Matthew Halm, "Molten Circulation and Rhetoric's Materiality," *Enculturation* 35 (2023), https://enculturation.net/molten_circulation
72. Byron Hawk, "A Diagrammatics of Persuasion," in *Circulation, Writing, and Rhetoric*, Laurie E. Gries and Collin Gifford Brooke, eds. (Utah State University Press, 2018), 308–314.
73. Pringle, Koch, and Stiegler, *Machine*, 74.
74. Raymond Williams, *Culture and Society: 1780–1950* (Columbia University Press, 1983).
75. Raymond Williams, *Keywords: A Vocabulary of Culture and Society* (Oxford University Press, 2015).
76. Joni Adamson, William Gleason, and David Pellow, *Keywords for Environmental Studies*, J. Adamson, W. A. Gleason, and D. Pellow, eds. (New York University Press, 2016). For example, at the time of this writing, the Living Lexicon section of the journal *Environmental Humanities* contains no entry for system or ecosystem, and it is also absent from *Keywords for Environmental Studies*.

77. Madison Jones, "A Counterhistory of Rhetorical Ecologies," *Rhetoric Society Quarterly* 54, no. 4 (2021), https://doi.org/10.1080/02773945.2021.1947517
78. Jenny Edbauer, "Unframing Models of Public Distribution: From Rhetorical Situation to Rhetorical Ecologies," *Rhetoric Society Quarterly* 35, no. 4 (2005): 5–24, doi:10.1080/02773940509391320. Following Edbauer's essay on public rhetoric and ecology, featuring a study of the "Keep Austin Weird" slogan, rhetorical ecologies have led to many studies focusing on place-based methods for conducting place-based research. Numerous recent articles, special issues of journals, and edited collections have drawn from ecology to suggest the importance of fieldwork in rhetorical criticism. See Samantha Senda-Cook et al., eds., *Readings in Rhetorical Fieldwork* (Routledge, 2019); Michael K. Middleton, Samantha Senda-Cook, and Danielle Endres, "Articulating Rhetorical Field Methods: Challenges and Tensions," *Western Journal of Communication* 75, no. 4 (2011): 386–406, doi:10.1080/10570314.2011.586969; Michael K. Middleton, Samantha Senda-Cook, and Danielle Endres, *Participatory Critical Rhetoric: Theoretical and Methodological Foundations* In Situ, (Rowman & Littlefield, 2015); Roberta Chevrette et al., "Rhetorical Field Methods/Rhetorical Ethnography," *Oxford Research Encyclopedia of Communication* (2023), https://doi.org/10.1093/acrefore/9780190228613.013.1378. Other studies have extended this work to focus on rhetorical ecologies and contextual fields. For example, see Bridie McGreavy et al., eds., *Tracing Rhetoric and Material Life Ecological Approaches* (Palgrave Macmillan, 2018); Kent A. Ono, "Contextual Fields of Rhetoric," *Western Journal of Communication* 84, no. 3 (2020): 264–279, doi:10.1080/10570314.2019.1681497. Other scholars have focused on field methods for cultivating participatory scholarship and pedagogy and conducting ethnographic research. See Danielle Endres, A. Hess., S. Senda-Cook, and M. K. Middleton, eds., "Rhetorical Fieldwork [Special Issue]." *Cultural Studies ↔ Critical Methodologies* 16, no. 6 (2016): 511–80. doi:10.1177/1532708616655820; Sara L. McKinnon et al., eds., *Text + Field: Innovations in Rhetorical Method* (Penn State University Press, 2017); Candice Rai and Caroline Gottschalk Druschke, eds., *Field Rhetoric: Ethnography, Ecology, and Engagement in the Places of Persuasion* (University of Alabama Press, 2018); C. G. Druschke, "A Trophic Future for Rhetorical Ecologies." *Enculturation: A Journal of Rhetoric, Writing, and Culture* 28, no. 1 (2019), http://enculturation.net/a-trophic-future; Phaedra C. Pezzullo and Catalina M. de Onís, "Rethinking Rhetorical Field Methods on a Precarious Planet," *Communication Monographs* 85, no. 1 (2018): 103–122, doi:10.1080/03637751.2017.1336780. Taken

together, these studies demonstrate the potential and the growing popularity of in situ (or place-based) work influenced by ecological science.
79. Caroline Gottschalk Druschke, "The Radical Insufficiency and Wily Possibilities of RSTEM," *POROI* 12, no. 2 (2017): 1–10, doi:10.13008/2151-2957.1257
80. Taylor, *Unruly Complexity Ecology, Interpretation, Engagement.*
81. Dana Phillips, *The Truth of Ecology: Nature, Culture, and Literature in America* (Oxford University Press, 2003).
82. See Patrick Kangas, *A History of Radioecology* (Routledge, 2022).
83. Laura J. Martin, *Wild by Design: The Rise of Ecological Restoration* (Harvard University Press, 2022). For an in-depth history of ecology, from conservation to restoration, including an in-depth discussion of H.T. Odum and his contemporaries, see Martin.
84. Caroline Gottschalk Druschke, "A Trophic Future for Rhetorical Ecologies," *Enculturation: A Journal of Rhetoric, Writing, and Culture* 28, no. 1 (2019): para. 5, http://enculturation.net/a-trophic-future
85. Linda Tuhiwai Smith, *Decolonizing Methodologies: Research and Indigenous Peoples* (Zed Books, 1999), 2.
86. Natasha N. Jones, Kristen R. Moore, and Rebecca Walton, "Disrupting the Past to Disrupt the Future: An Antenarrative of Technical Communication," *Technical Communication Quarterly* 25, no. 4 (2016): 212, doi:10.1080/10572252.2016.1224655; Smith, *Decolonizing Methodologies*, 2. In practice, field histories follow Tuhiwai Smith's concept of history, which resists or undercuts Eurocentric narratives, emphasizing "the history of Western research through the eyes of the colonized."
87. Aja Y. Martinez, *Counterstory: The Rhetoric and Writing of Critical Race Theory* (National Council of Teachers of English, 2020), 5.
88. Ibid., 58.
89. April L. O'Brien and James Chase Sanchez, *Countermemory: A Rhetoric of Resistance* (University of Alabama Press, 2025). Field histories follow the work of April L. O'Brien and James Chase Sanchez on countermemory, which they define as a mixed-methods approach to rhetoric scholarship that brings together place-based analysis with historiography to resituate the narratives of marginalized groups.
90. Ryan Skinnell, "Who Cares If Rhetoricians Landed on the Moon? Or, a Plea for Reviving the Politics of Historiography," *Rhetoric Review* 34, no. 2 (2015): 113, doi:10.1080/07350198.2015.1008907
91. Druschke, "A Tropic Future," para. 39.
92. Danielle Endres, "The Most Nuclear-Bombed Place: Ecological Implications of the US Nuclear Testing Program," in *Tracing Rhetoric and Material Life: Ecological Approaches*," Bridie McGreavy, et al., eds. (Palgrave Macmillan, 2018). For more information on nuclear colonialism in the American West, specifically as it pertains to more-than-human

rhetorics in the Anthropocene, see Danielle Endres' "The Most Nuclear-Bombed Place: Ecological Implications of the US Nuclear Testing Program."
93. Laura J. Martin, "Proving Grounds: Ecological Fieldwork in the Pacific and the Materialization of Ecosystems," *Environmental History* 23, no. 3 (2018): 1, doi:10.1093/envhis/emy007
94. Patrick Kangas, "The Role of Passive Electrical Analogs in H.T. Odum's Systems Thinking," *Ecological Modelling* 178, no. 1–2 (2004): 7, doi:10.1016/j.ecolmodel.2003.12.019
95. Joel Hagen, *An Entangled Bank: The Origins of Ecosystems Ecology* (Rutgers University Press, 1992), 101.
96. Kenneth Burke, *A Grammar of Motives* (University of California Press, 1969), 73.
97. Howard Washington Odum, *Southern Regions of the United States* (University of North Carolina Press, 1936).
98. Ariel E. Lugo, "H.T. Odum and the Luquillo Experimental Forest," *Ecological Modelling*, 178 (2004): 68, doi:10.1016/j.ecolmodel. 2003.12.023
99. Mark Brown, "HT Odum #5 ARIEL LUGO," podcast, in *The Legacy of H.T. Odum and System Ecology*, YouTube, https://www.youtube.com/watch?v=m3oshBegmgs
100. Ibid.; Lugo, "H.T. Odum," 69–71. See Lugo, 69–71.
101. Gordon H. Orians, "A Diversity of Textbooks: Ecology Comes of Age," *Science* (1973): 1238, doi:10.1126/science.181.4106.1238
102. Skinnell, "Who Cares If," 113.
103. Ibid., 115.
104. Jennifer Clary-Lemon, "Gifts, Ancestors, and Relations: Notes toward an Indigenous New Materialism," *Enculturation: A Journal of Rhetoric, Writing, and Culture* 30, no. 1 (2019): para. 17, http://enculturation.net/gifts_ancestors_and_relations
105. Lydia Wilkes, "'Becoming Daibook': Avowing Settlerness to Reduce Settler Harm in Rhetoric, Composition, and Writing Studies," *College Composition & Communication* 76, no. 2 (2024): 312, doi:10.58680/ccc2024762310
106. Simone Bignall and Daryle Rigney, "Indigeneity, Posthumanism and Nomad Thought Transforming Colonial Ecologies," in *Posthuman Ecologies: Complexity and Process after Deleuze*, Rosi Braidotti and Simone Bignall, eds. (Rowman & Littlefield, 2018), 159.
107. Ibid., 160.
108. Clary-Lemon, "Gifts, Ancestors, and Relations," para. 3.
109. Kristin Arola, "A Land-Based Digital Design Rhetoric," in *Routledge Companion to Digital Writing & Rhetoric*, Jonathan Alexander and Jacqueline Rhodes, eds. (New York: Routledge, 2018).

110. David M. Grant, "Writing Wakan: The Lakota Pipe as Rhetorical Object," *College Composition & Communication* 69, no. 1 (2017): 61–86. doi:10.58680/ccc201729296
111. David M. Grant, "Like Frost on a Windowpane: On the Pluriversal Possibilities of Spacetime," *Enculturation: A Journal of Rhetoric, Writing, and Culture* 31, no. 1 (2020), http://enculturation.net/like_frost
112. Donnie Johnson Sackey et al., "Perspectives on Cultural and Posthumanist Rhetorics," *Rhetoric Review* 38, no. 4 (2019): 375–401, doi:10.108 0/07350198.2019.1654760
113. Clary-Lemon, "Gifts, Ancestors, and Relations," para. 8.
114. Kenneth Burke, *A Grammar of Motives* (University of California Press, 1969), xvi-xvii.
115. Kenneth Burke, *Attitudes Toward History* (University of California Press, 1937), 150. See also Marika Seigel, "'One Little Fellow Named Ecology': Ecological Rhetoric in Kenneth Burke's Attitudes toward History." *Rhetoric Review* 23 (2004): 388–403, https://doi.org/10.1207/s15327981rr2304_6
116. Laurence Coupe, "Kenneth Burke: Pioneer of Ecocriticism," *Journal of American Studies* 35, no. 3 (2001): 413–431, https://doi.org/10.1017/S0021875801006697
117. Robert Wess, "Ecocriticism and Kenneth Burke: An Introduction," *K. B. Journal* 2, no. 2 (2006), https://www.kbjournal.org/wess2
118. Richard M. Coe, "Rhetoric 2001," *Freshman English News* 3, no. 1 (1974): 1–3; 9–13, https://www.jstor.org/stable/43518660
119. Richard M. Coe, "Eco-Logic for the Composition Classroom," *College Composition and Communication* 26, no. 3 (1975): 232–237, doi:10.2307/356121
120. Richard M. Coe, "Closed System Composition," ETC., *A Review of General Semantics* 32, no. 4 (1975): 403–412, https://www.jstor.org/stable/42582292
121. Robert L. Scott, "On Not Defining 'Rhetoric,'" *Philosophy & Rhetoric* 6, no. 2 (1973): 81–96, https://www.jstor.org/stable/40236837
122. Marilyn Cooper, "The Ecology of Writing," *College English* 48, no. 4 (1986): 364–375, doi:10.2307/377264
123. Karen Burke LeFevre, *Invention as a Social Act* (Southern Illinois University Press, 1987).
124. Byron Hawk, "Counter-Traditions in Ecologies of Composition: Three Models of Futurity," in *Rhetorical Ecologies*, Sidney I. Dobrin and Madison Jones, eds. (NCTE Press, 2024), 39.
125. Louise Wetherbee Phelps, *Composition as a Human Science: Contributions to the Self-Understanding of a Discipline* (Oxford University Press, 1988).
126. Margaret A. Syverson, *The Wealth of Reality: An Ecology of Composition* (Southern Illinois University Press, 1999).

127. Edbauer, "Unframing Models," 5–24.
128. Byron Hawk, *A Counter-History of Composition: Toward Methodologies of Complexity* (University of Pittsburgh Press, 2007), 40.
129. Dan Ehrenfeld, "'Sharing a World with Others': Rhetoric's Ecological Turn and the Transformation of the Networked Public Sphere," *Rhetoric Society Quarterly* 50, no. 5 (2020): 305–320, doi:10.1080/02773945.2020.1813321
130. Bridie McGreavy et al., eds., *Tracing Rhetoric and Material Life Ecological Approaches* (Palgrave Macmillan, 2017).
131. Tyler S. Rife, "(Dis)entangling the ontopolitics of ecological theorizing for critical communication," *Review of Communication* 24, no. 4 (2024): 247–259, doi: 10.1080/15358593.2024.2413070
132. Linli Lan and Ying Yuan, "Ecological Rhetoric: Strands and Trends," *Linguistics and Literature Studies* 10, no. 5 (2022): 85–94, doi: 10.13189/lls.2022.100501
133. Hawk, *A Counter-History*.
134. Dobrin and Jones, *Rhetorical Ecologies*.
135. Dobrin and Weisser, *Ecocomposition: Theoretical*.
136. Nathaniel A. Rivers and Ryan P. Weber, "Ecological, Pedagogical, Public Rhetoric," *College Composition and Communication* 63, no. 2 (2011): 187–218, doi:10.58680/ccc201118389
137. John McDaid, "Toward an Ecology of Hypermedia," in *Evolving Perspectives on Computers and Composition Studies: Questions for the 1990s*, Gail Hawisher and Cynthia L. Selfe, eds. (National Council of Teachers of English, 1991), 203–223.
138. Colin Gifford Brooke, *Lingua Fracta: Toward a Rhetoric of New Media* (Hampton Press, 2009).
139. Sidney I. Dobrin and Sean Morey, *Ecosee: Image, Rhetoric, Nature* (SUNY Press, 2009).
140. Nedra Reynolds, *Geographies of Writing: Inhabiting Places and Encountering Difference* (Southern Illinois Press, 2004).
141. Anis Bawarshi, "The Ecology of Genre," in *Ecocomposition: Theoretical and Pedagogical Approaches*, Sidney I. Dobrin and Christian R. Weisser, eds. (State University of New York Press, 2001), 69–80.
142. Clay Spinuzzi, *Tracing Genres through Organizations: A Sociocultural Approach to Information Design* (MIT Press, 2003).
143. Stuart Blythe, "Agencies, Ecologies, and the Mundane Artifacts in Our Midst," in *Labor, Writing Technologies, and the Shaping of Composition in the Academy*, Pamela Takayoshi and Patricia Sullivan, eds. (Hampton Press, 2006), 167–186.
144. Hanna J. Rule, *Situating Writing Processes* (WAC Clearinghouse; University Press of Colorado, 2019).

145. Asao B. Inoue, *Antiracist Writing Assessment Ecologies: Teaching and Assessing Writing for a Socially Just Future* (The WAC Cleaninghouse; Parlor, 2015).
146. Michael Day, "The Administrator as Technorhetorician: Sustainable Technological Ecologies in Writing Programs," in *Technological Ecologies and Sustainability*, ed. Dànielle Nicole DeVoss, Heidi A. McKee, and Richard Selfe (Computers and Composition Digital Press; Utah State University Press, 2009), 130–148, http://ccdigitalpress.org/tes
147. Laurie Gries and C. G. Brooke, *Circulation, Writing, and Rhetoric* (Utah State University Press, 2018).
148. Kristie S. Fleckenstein, Clay Spinuzzi, Rebecca J. Rickly, and Carole Clark Papper, "The Importance of Harmony: An Ecological Metaphor for Writing Research," *College Composition & Communication* (2008), doi:10.58680/ccc20086871
149. James J. Brown, Jr., "Louis C. K.'s 'Weird Ethic': Kairos and Rhetoric in the Network," *Present Tense* 3, no. 1 (2013), https://www.presenttensejournal.org/volume-3/louie-c-k-s-weird-ethic-kairos-and-rhetoric-in-the-network/
150. Douglass Eyman, *Digital Rhetoric Theory, Method, Practice* (University of Michigan Press, 2015).
151. Noah Patton and Rachel Presley, "Twenty Years of Community Building: *Reflections* on/and Rhetorical Ecologies," *Reflections* 20, no. 1 (2020): 193–212, https://reflectionsjournal.net/wp-content/uploads/2020/09/V20.N1.Patton.pdf
152. Dànielle Nicole DeVoss, Heidi A. McKee, and Richard Selfe, eds., *Technological Ecologies & Sustainability* (Computers and Composition Digital Press; Utah State University Press, 2009), para. 2.
153. Chris Mays, "Writing Complexity, One Stability at a Time: Teaching Writing as a Complex System," *College Composition & Communication* 68, no. 3 (2017): 559–585, doi:10.58680/ccc201728966
154. Chris Mays, *Invisible Effects: Rethinking Writing through Emergency* (Peter Lang, 2021).
155. Druschke, "A Trophic Future."
156. Rai and Druschke, *Field Rhetoric*.
157. Robin E. Jensen, "An Ecological Turn in Rhetoric of Health Scholarship: Attending to the Historical Flow and Percolation of Ideas, Assumptions, and Arguments," *Communication Quarterly* 63, no. 5 (2015): 522–526, doi:10.1080/01463373.2015.1103600
158. Rich Shivener and Dustin Edwards, "The Environmental Unconscious of Digital Composing: Mapping Climate Change Rhetorics in Data Center Ecologies," *Enculturation* 32 (2020), https://enculturation.net/environmental_unconscious

159. Tim Jensen, *Ecologies of Guilt in Environmental Rhetorics* (Palgrave, 2019), doi:10.1007/978-3-030-05651-3
160. Nathan Stormer and Bridie McGreavy, "Thinking Ecologically about Rhetoric's Ontology: Capacity, Vulnerability, and Resilience," *Philosophy & Rhetoric* 50, no. 1 (2007): 1–25, doi:10.5325/philrhet.50.1.0001
161. Matthew Ortoleva, "Let's Not Forget Ecological Literacy," *Literature in Composition Studies* 1, no. 2 (2013): 66–73, doi:10.21623/1.1.2.5
162. Gabriela Raquel Rìos, "Cultivating Land-Based Literacies and Rhetorics," *Literacy in Composition Studies* 3, no. 1 (2015): 60–70, doi:10.21623/1.3.1.4
163. Stephanie J. West-Puckett, "Decentering Whiteness in the Digital Humanities Through/With Decolonial Methodologies ~ Session F7," *Digital Rhetoric Collaborative*, July 17, 2014, para. 2, https://www.digitalrhetoriccollaborative.org/2014/07/17/decentering-whiteness-in-the-digital-humanities-throughwith-decolonial-methodologies-session-f7/
164. Ibid., para. 9.
165. Louis M. Maraj, *Black or Right: Anti/Racist Campus Rhetorics* (University Press of Colorado, 2020), 7.
166. Dustin W. Edwards, "Critical Infrastructure Literacies and/as Ways of Relating in Big Data Ecologies," *Computers and Composition* 61 (2021): 5, doi:10.1016/j.compcom.2021.102653
167. McGreavy et al., *Tracing Rhetoric*, 8, 12. While it is beyond the scope of this chapter to fully engage with the vast and overwhelming body of scholarship that is resonant with ecology across the entirety of RWCS, McGreavy and her co-editors offer a compelling history of its development in environmental communication in the introduction to *Tracing Rhetoric and Material Life: Ecological Approaches*. They point to its origins in constitutive rhetorics in the 1970s and 1980s, where communication scholars drew from Burke to "forward 'materialist' accounts of rhetoric." They offer a compelling history of the concept from this origin point into "transhumanist and articulation models of constitutive communication."
168. Noah Roderick, "Analogize This! The Politics of Scale and the Problem of Substance in Complexity-Based Composition," *Composition Forum* 25, (2012): par. 1, http://compositionforum.com/issue/25/scale-substance-complexity.php
169. Ehrenfeld, "Sharing a World," 4.
170. Donna Haraway, *Staying with the Trouble: Making Kin in the Chthulucene* (Duke University Press, 2016): 1.

171. In demonstrating this spatiotemporal violence, I build from Anibal Quijano's argument that "Europeans generated a new temporal perspective of history and relocated the colonized population, along with their respective histories and cultures, in the past of a historical trajectory whose culmination was Europe" (541, qtd. in Clary-Lemon).
172. Félix Guattari, "Machinic Heterogenesis," in *Chaosmosis: An Ethico-Aesthetic Paradigm*, trans. Paul Bains and Julian Pefanis (Indiana University Press, 1995), 19–20.
173. Martin, "Proving Grounds," 3.
174. Miller, "*Kairos*."
175. Martin, "Proving Grounds," 579.
176. Megan Eatman, *Ecologies of Harm: Rhetorics of Violence in the United States* (Ohio State University Press, 2020), 2.
177. Jack Niedenthal, "A History of the People of Bikini Following Nuclear Weapons Testing in the Marshall Islands: With Recollections and Views of Elders of Bikini Atoll," *Health Physics* 73, no. 1 (1997): 28.
178. Leah Ceccarelli, *On the Frontier of Science: An American Rhetoric of Exploration and Exploitation* (Michigan State University Press, 2013), 3.
179. Max Liboron, *Pollution Is Colonialism* (Duke University Press, 2021).
180. Danielle Endres, *Nuclear Decolonization: Indigenous Resistance to High Level Nuclear Waste Siting* (Ohio State University Press, 2023), ix.
181. Ibid., ix.
182. Liboiron, *Pollution Is Colonialism*, 26.
183. Ibid., 6.
184. Ibid.
185. Ibid., 135.
186. Ibid., 6.

Open Access This chapter is licensed under the terms of the Creative Commons Attribution-NonCommercial-NoDerivatives 4.0 International License (http://creativecommons.org/licenses/by-nc-nd/4.0/), which permits any noncommercial use, sharing, distribution and reproduction in any medium or format, as long as you give appropriate credit to the original author(s) and the source, provide a link to the Creative Commons license and indicate if you modified the licensed material. You do not have permission under this license to share adapted material derived from this chapter or parts of it.

The images or other third party material in this chapter are included in the chapter's Creative Commons license, unless indicated otherwise in a credit line to the material. If material is not included in the chapter's Creative Commons license and your intended use is not permitted by statutory regulation or exceeds the permitted use, you will need to obtain permission directly from the copyright holder.

CHAPTER 2

Energy Rhetoric in Odum's "Silver Springs Study"

Abstract Turning to one of H. T. Odum's most influential ecological studies, the "Silver Springs study," this chapter examines the rhetoric of energy in ecosystems ecology. Energy flow became a dominant metaphor for ecosystems ecology, shaping the way scientists and the public imagine relationships within environments. This chapter analyzes the role of cybernetics, thermodynamics, and economics in the invention of an ecological framework that emphasized efficiency, control, and optimization. These values persist in shaping contemporary ecological inquiry.

Keywords Energy flow • H. T. odum • Cybernetics • Thermodynamics • Ecological modeling • Metaphor in science

> Behold, for instance, a vast circular expanse before you, the waters of which are so extremely clear as to be absolutely diaphanous or transparent as the ether. […] This amazing and delightful scene, though real, appears at first but as a piece of excellent painting; there seems no medium, you imagine the picture to be within a few inches of your eyes, and that you may without the least difficulty touch any one of the fish, or put your finger upon the crocodile's eye, when it really is twenty or thirty feet under water. —William Bartram, *Travels*[1]

© The Author(s) 2026
M. P. Jones, *Inventing Ecosystems*, Palgrave Studies in Media and Environmental Communication,
https://doi.org/10.1007/978-3-031-98793-9_2

Silver Springs State Park, Silver Springs, Florida, United States
I run my paddle through the crystal water, marveling at how the wake moves like molten glass, distorting the visual field where I notice outlines of turtles, damselfly nymphs, and numerous fish shimmering through the vacillating ribbons of eelgrass far below where my boat dithers against the tranquil surface. Just out of reach, a cormorant perches on a log, drying his wings beside a group of red-eared sliders, aligned smallest to largest and sunning themselves in the warm spring afternoon light. I picture the words of American naturalist, writer, and explorer, William Bartram, who remarked on the clarity of the water during his 1774 visit to nearby Salt Springs in his famous book Travels. *In that passage, Bartram imagines the lucid water as fostering a kind of equality in nature, a "paradise of fish" among the many species who inhabit the spring.*[2] *He attributes this more egalitarian regime to the clear spring water, which provides the fish with an open viewing pane that makes predation more difficult. Today, the concept of the "ecology of fear" describes how predator–prey relations shape environments through affect—the terror experienced by prey—in addition to physical population control. It is little wonder that these crystal springs—where herons stalk the pickerelweed and bees and dragonflies hum around the spider lily, where rhesus monkeys hide in dark forests of live oak draped in Spanish moss, the air is filled with insect, frog, and birdsong, and where coral snakes, cottonmouths, and alligators slither through floors paved with cypress knots and ornamented with palmetto spears—could be seen to stand in for the very Platonic ideal of "Nature" itself.*

Silver Springs and the Ecosystem

This visit to Silver Springs State Park in 2016 was my introduction to the ways that Florida's springs have long shaped environmental thinking. In fact, Florida's freshwater springs have inspired budding American environmental imaginations for hundreds of years, as Bartram's writings would influence British Romantics like S.T. Coleridge and William Wordsworth,[3] American Nature Writing through figures like R.W. Emmerson and H.D. Thoreau,[4] as well as the African American Literary Tradition through writers like Zora Neale Hurston.[5] Silver Springs was also a tourist destination before the Civil War, and its popularity grew with the invention of the glass bottom boat tour there in the late 1870s, which provided a new technology through which to view the natural world. As Wendy Adams King puts it, "…[u]pon the Silver River's waters and within the glass-bottom boat, the social significance of America and its landscape is negotiated within tensions among romantic, scientific, and cinematic visions."[6] These tensions are also present in the films and television programs that were shot there, such as *Sea Hunt*, the James Bond film *Moonraker*, the 1930s and 1940s Tarzan movie series, and *Creature from the Black Lagoon*. Silver Springs is also home to Fort King, which was an important location during the Second Seminole War, including the infamous Dade Massacre, and was part of the large-scale colonial project of genocide and removal of Indigenous peoples that was used in part to produce Romantic views of "Nature," a subject that we will return to in the next chapter. King connects this visioning of the natural world as a frame made of "Western myth and Renaissance, Enlightenment, and Romantic ideals."[7] The seemingly self-contained dimensions of the Florida springs make them particularly strong candidates to represent a vision of Nature as distinct, and even isolated, from the human world.

Ecology and nature writing both share roots in natural history in the various forms it took across the eighteenth and nineteenth centuries. In fact, Tansley once remarked that apostates of ecology often view the discipline as "the old natural history masquerading under a high-sounding name-and not always very good natural history at that!"[8] The longstanding racism associated with multiple intellectual traditions in natural history—which sometimes applied human-centered and trophic hierarchies to social and racial categories—persists in shaping and limiting ecological inquiry and contributes to embedded racialized disciplinary environments today.[9] This inheritance shows up not only in the overt affiliations between the origins of ecology with colonialism, racism, and eugenics—evident in examples like Ernst Haeckel's enthusiastic support of Social Darwinism—but also in the implicit agonism of ecological concepts like holism and apartheid (Smuts), genes and behavior, and (mis)interpretations of fitness. For example, proto-ecological thought in the early eighteenth century is often organized into the opposing camps of Arcadian/romantic/preservationist (associated with figures like Gilbert White) and Imperial/managerial/pastoralist (associated with figures like Francis Bacon),[10] though Peder Anker shows these distinctions to be generally "false and anachronistic dichotomies" that are better understood as a "tangled web of both imperial and romantic views."[11] To oversimplify, Arcadian thinking sought simplicity and harmony between humans and nature while Imperialism sought dominance and control through the capacity of reason. There exists agonism among these traditions, and these conflicts crop up throughout ecology's origins and development. Bartram's work and its wide-ranging influence exemplifies the development of the American ecological tradition from naturalism, which holds subtle distinctions from its development on other continents.[12] Among these divergent practices, there exists a tension between preserving natural environments and controlling their productivity.[13] The work of figures like Carl Linnaeus, Alexander von Humboldt, and Charles Darwin helped to propagate the disciplinary shift from what Peter Ayres refers to as "the purely descriptive approaches of the eighteenth century" to Haeckel's conception of *Oekologie* as its own discipline.[14] Haeckel attributes Darwin's work as setting up the model of mechanistic relations between organisms and environments that underpin ecology.[15] These mechanical views of nature reproduced Arcadian and Imperialist agonisms between interfacing with environments and controlling them.

Such frames, tensions, and influences certainly helped shape Howard T. Odum's famous Silver Springs study, which ran from 1951 until 1956. In this watershed study, Odum first mapped the trophic structure and energy flows of the springs, inaugurating an important moment both in Odum's career and in ecological science more generally. This study provided the first comprehensive analysis of a natural ecosystem, bringing controlled field experiments together with extensive data collection and cutting-edge quantitative methods to understand, model, and visualize productivity and/as ecological power output moving through a closed system.[16] This research not only advanced the field of ecosystems ecology by emphasizing the importance of energy circulation but also laid the groundwork for understanding the springs as both a natural resource and site of significant ecological and rhetorical value. The crystal clear water, which flowed from underground at a steady rate, provided more than a visual representation of Romantic Nature. As Odum explains, the springs function "collectively [as] a giant constant temperature laboratory" providing a "rare situation [in which] it is possible to compare whole communities in a ready made experimental design."[17] The steady flow of the spring's water provided a measurable and observable means to map these energy exchanges, making Silver Springs a key site for ecological study and environmental education. The stable environment provided the perfect conditions for applying systems theory to study the natural world. Today, as the springs are impacted by a host of anthropogenic environmental problems, they no longer offer the same closed system. As such, Odum's study also serves as a baseline of water quality and productivity along with a follow-up study in the 1980s by his doctoral student Robert L. Knight, who examined the metabolism, productivity, and consumer control structure of the springs.[18] In 2006, these and other studies were compiled into the comprehensive *Fifty-Year Retrospective Study of the Ecology of Silver Springs*, which documented changes in the ecology of the springs.[19] Since Odum's initial study, changing environments have drastically altered the ecology and hydrology of Florida's springs.

As I paddle along the Silver River, I notice *Lyngbya*, a type of invasive algae, overtaking a native red *Ludwigia* plant along the river bottom. From high above, the algae resembles a massive cloud of green dust, swallowing everything in its path. My thoughts drift to the many different types of threats facing the ecosystem, from an increase in nutrients in the water, which feed algal growth, to a reduction in discharge rate, caused by changing rainfall patterns and massive groundwater withdrawals, to saltwater intrusion from rising sea levels, to increasing numbers of visitors to the springs, to loss of habitat caused by invasive species. This thought gives me pause and calls for me to reflect on the work that Caroline Gottschalk Druschke and her coauthors have done to investigate the ways that such species are "frequently characterized through xenophobic, militaristic, monstrous, and disease metaphors" and "those metaphors—and language choices more generally—had significant positive and negative consequences for science, management, and the world at large."[20] Rather, the disruptions in ecological systems can be viewed as indicators of the larger health of an ecological community.

Along similar lines, Donnie Johnson Sackey argues in *Trespassing Natures: Species Migration and the Right to Space* that the massive scale of anthropogenic environmental changes calls for "a new paradigm," one that can "attend to the social factors that structure beliefs about who does or does not belong and abandon the belief that species invade."[21] Sackey unearths the history of the "invasive" metaphor, and he details the ways that place and time constrain views of "flora and fauna [that] are static to the point that we can establish an idea of nativity."[22] Odum's research, starting with the Silver Springs study, helped establish disturbance ecology as an important research area within ecology, viewing changes to ecosystems over various spatial and temporal scales. As I will demonstrate in Chap. 3, as well as elaborate in Chaps. 4 and 5, these spatial and temporal frameworks are deeply rooted in the colonial history of ecosystems ecology. As indicators of the health of a closed ecosystem, the changes in the number and distribution of different species in the springs have more to show us about human impacts than they do about a particular species of algae. For instance, the excess algal growth is directly fueled by an increase of nutrients in the water, namely nitrogen and phosphorus, which are linked to environmental disturbance caused by development, as well as excessive fertilizer use and failing septic systems. Odum's Silver Springs study helped record an extensive snapshot of the ecology of the area, and the follow-up studies offer important evidence of its decline and need for protection.

Energy Systems

Just as Raymond Lindeman was, in part, inspired to imagine the ecosystem by Freud's conception of the human brain as an electrical network, so can the excess nutrients in the water help us to imagine the health of the springs at the scale of energy flow. H. T. Odum developed *energese*, or an "energy systems language," working with colleagues in El Yunque in Puerto Rico (a topic I will return to in the next chapter). Energese is a modeling language that can be used to produce diagrams explaining how systems function by reducing trophic dynamics to abstract symbols (Fig. 2.1).

The resulting diagrams resemble an electrical circuit more than they might an environment. Furthermore, the reduction of environments to an energy system language is one of the places where technocracy's influence on ecosystems ecology is most apparent. As Taylor explains, technocrats sought to replace not only democracy but also capitalism, especially the "price system," with "equal allocations of nonaccumulable energy certificates."[23] This focus on energy as the central metric and mechanism for societal control was influential in shaping G. Evelyn Hutchinson's thinking about ecosystems, which was more fully realized in the work of H. T. Odum, who was his doctoral student at Yale. Taylor details the history of the Macy Conferences, an interdisciplinary series of meetings on cybernetics that ran from 1941 to 1960, in which Hutchinson participated. The role of communications in cybernetics became an important part of the Macy conferences following WWII, where the systems thinking espoused by the group began to have "technocratic implications."[24] Taylor demonstrates that Hutchinson's "systems approach to understanding nature moved easily into a systems approach for engineering society."[25] Thus, the energy rhetoric of the technocracy movement found purchase in H. T. Odum's thinking, who had been fascinated from a young age by

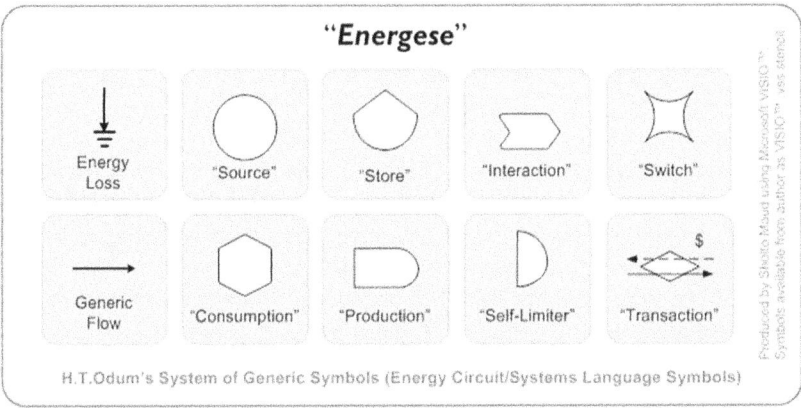

Fig. 2.1 Sample visualizations of H. T. Odum's "System of Generic Symbols," adapted with permission from Sholto Maud's stencils, via Wikimedia Commons, https://commons.wikimedia.org/wiki/File:Energy_Systems_Symbols_H.gif (Public Domain). This iconographic approach to ecosystem modeling was introduced by H. T. Odum and Elisabeth C. Odum in *Modeling for All Scales: An Introduction to System Simulation* (2000)

electrical circuits,[26] which he learned about from Alfred Morgan's *The Boy Electrician*.[27]

Energese makes possible the translation of a closed system like Silver Springs into an iconographic series of pictograms, depicting the movement of energy through the various parts of an ecosystem. From this perspective, the algal growth has as much to do with the excess energy in the system (phosphorus and nitrogen) as it does with the metaphors of invasion that both Sackey and Druschke et al. critique. These factors are evidence of larger ecosystem disturbance, rather than simply the intrusion of an individual species on a pristine place. At the same time, trophic mapping also relies on what I refer to as a *spatiotemporal slice*, which I will discuss further throughout Chaps. 3, 4, and 5, producing a snapshot of the ecosystem in a particular place and at a particular moment in time. By comparing the initial study in 1957 with subsequent studies in 1980 and 2006, the degradation of the springs becomes visible at a larger scale than a single invasive species. Elsewhere, I have discussed the ways that the degradation in clarity has been used as a trope in networked activism projects that bring together art, science, research, and advocacy to help protect Florida's iconic springs.[28] In that study, I argue that, as part of a longstanding tradition of visual rhetoric in environmental politics, these organizations draw "upon the socio-affective circulation [...] of springs iconography" in order to "deploy a fluvial rhetoric which engages with [the energy of] place."[29] Building from that work, this chapter seeks to explore how Silver Springs is a site that makes visible the early links between Odum's research in trophic dynamics, his later application of energese to study ecosystems, and the troubles that this theoretical basis produces in contemporary ecological inquiry. As we import systems thinking into DH/EH, and into RWCS more specifically, we risk importing the rhetoric of technocratic optimism that characterized early ecosystems ecology. As such, it is important to briefly unpack the ways that energy might be characterized as rhetorical, as well as the ways that rhetoric could be said to be energetic.

Energy and/as Rhetoric

Energy holds deep conceptual ties to the discipline of rhetoric going back at least to antiquity. In "Energy: Rhetoric's Vitality," Chris Ingraham artfully traces the tangled conceptual and etymological threads that together weave both the rhetoric *of* energy and rhetoric *as* energy.[30] While Ingraham reveals the complex and convoluted nature of energy in rhetoric, he also offers useful ways of mapping this capacious concept onto contemporary rhetorical theory. Similarly, Michael Marder's *Energy Dreams: Of Actuality* offers a philosophical account of the "relative and absolute ambiguities of energy," which add layers of confusion to the "historical predicament of energy today."[31] The predicament he describes, energy's ambiguity, begins with Aristotle and twists its way through premodern European theory to nineteenth-century science, to find ubiquitous application in contemporary discourse, from popular culture to technical science. In modern rhetorical terms, "energy" connotes vigorous or vivid expression. Aristotle distinguishes between energy in *Metaphysics* in terms of potentiality (or *dunamis*) and actuality (or *energeia*). Marder explains how the relationship between potential energy and actualized energy immediately creates confusion, as the modern usage of energy "is the inverse of Aristotle's."[32] To make matters worse, Ingraham explains how the paradigm shifts between "Newtonian mechanics" and "Einstein's theory of special relativity [...] pivots on fundamentally different conceptions of energy."[33] As Lindeman developed his ecosystem metaphor out of Freud's interest in the mind as an electrical network, he invoked this tangled rhetorical tradition between different types of energy and the ways that energy is seen as potential (or stored) and actual (or kinetic).

For our purposes, consider the rhetorical work of energy in Odum's Silver Springs study, in which he examines how light enters the spring pool where it joins a steady flow of nutrients and other inputs, which are absorbed and metabolized by a wide variety of organisms.[34] In this model, we might ask if it is the light (and other inputs) that holds the potentiality (*dunamis*) to energize the ecosystem, or instead if it is the ecosystem that holds the potential to activate (*energeia*) the potential energy of the light (the spring system's capacity). This tautological knot is at the center of H. T. Odum's concept of productivity, which governs the ways that a "community metabolism is self-regulated."[35] He would go on to develop a "network language" that could represent energy transformations and transfers across different scales with the goal of demonstrating how complex symbiotic systems achieve stability over time.[36] As Patrick Kangas demonstrates, Odum's early work in the 1950s and 1960s with "simple electrical networks" laid the foundation for "the development of Odum's approach to systems."[37] Kangas points out that while Odum's "Ohm's Law analogy" for ecosystems was met with immediate criticism after publication, "Odum modified the analogy to address these criticisms and continued to use passive analogs as ecological models for another decade."[38] He demonstrates that these electrical models were one of the important strains of early work that helped produce modern ecosystems ecology.

The capaciousness of energy presents mirrored problems for both ecology and rhetoric, as the application of different types of energy, in all its potentiality and actuality, creates barriers to general theories of ecosystem, ecology, or rhetoric. In *Rhetorical Ecologies*, Sid Dobrin and I discuss, and ultimately trouble, the notion of a general rhetorical ecology, building from Erich Hörl's conception of a "general ecology."[39] While George A. Kennedy's 1992 article "A Hoot in the Dark: The Evolution of General Rhetoric" argues for a general theory of rhetoric as energy, his engagement with the problems of energy ultimately faces the same problems that energy poses for ecosystems ecology.[40] Odum's work was criticized for viewing energy as a "one dimensional" or "universal currency in ecology."[41] Such a focus seeks to reduce organism functions to those of machines or circuits, but ultimately fails because "the types of energy and their spatio-temporal partitioning in real ecosystems are too diverse."[42] In other words, energy is too ambiguous and broad to serve as a circulating medium for the ecosystem. This holds true for rhetoric as well, where energy, and its circulation, might serve as too convenient a term for the transmission and transformation of messages. As Ingraham points out, many scholars in contemporary rhetorical inquiry follow similar lines of thinking as Kennedy's article lays out, demonstrating that his essay has introduced new directions to the field.[43] In its most basic sense, Kennedy's use of energy sufficiently distorts thermodynamics such that it may serve to further obfuscate, rather than clarify, this new paradigm for rhetoric. While energy, like ecology, offers convenient purchase as a metaphor for RWCS, without care it may also support technocratic perspectives.

CIRCULATION, EMERGY, AND THE RHEME

Throughout the Odum brothers' careers, but especially H. T.'s later work with energese and the production of an energy systems language, the Odums built from Alfred J. Lotka's idea of the "maximum power principle," which attempted to explain Darwinian evolution at the scale of ecosystems and in terms of energy. As H. T. Odum puts it, "…over time a network that draws more resources and uses them better toward maintaining that network will tend to replace designs that have fewer resources with which to work."[44] Noting the problems posed by different scales of size and time (which I return to in Chaps. 4 and 5, respectively), he proposes "emergy"—a portmanteau of "embodied" and "energy"—as a "scale-independent measure of work and a useful concept of value."[45] Emergy—along with other units such as "emjoule" and "emcalorie"—was first suggested to Odum as a way to combine embodied energy by David M. Scienceman during a visiting appointment at the University of Florida in 1986.[46] Scienceman is an Australian scientist who changed his name from David Slade in 1972 as part of his work to establish a political movement and party representing science.[47] Scienceman also suggests additional emergy nomenclature such as "empower, emdollar, embit, energy memory and the maximum empower principle."[48] Starting in the mid-1980s, emergy emerges as the basic currency of Odum's ecosystem. He defines emergy as referring to the holistic amount of energy consumed in the work that supports a system. Odum gives the example of energy moving "from dilute sunlight up to plant matter, to coal, from coal to oil, to electricity and up to the high quality efforts of computer and human information processing."[49] Emergy allows Odum to trace the movement of energy through a closed system, be it a cypress dome, the production of information, a coral reef, or geopolitical conflict in Afghanistan (Fig. 2.2).

These diagrams may evoke for RWCS scholars (or at least those interested in RNM) Bruno Latour's actor-network theory (ANT), which

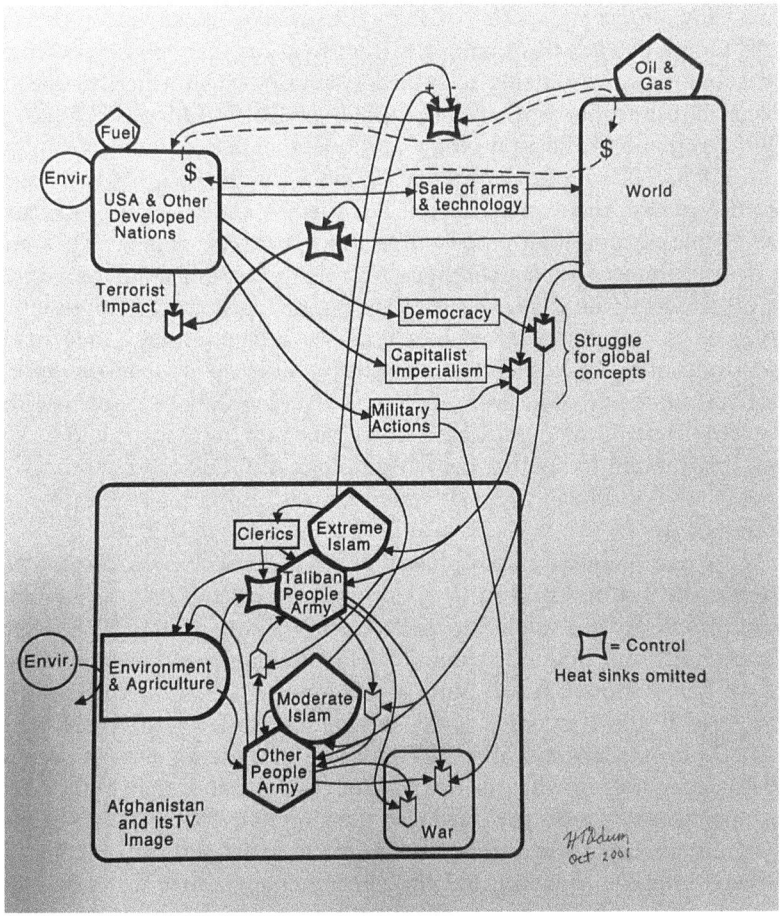

Fig. 2.2 An example of H. T. Odum's energy diagrams, depicting conflict in Afghanistan and the relationships between energy, war, religion, ideology, and media representation. Courtesy of Special & Area Studies Collections at the University of Florida

similarly maps relations and agency across both nonhuman and human phenomena. As Eugene Odum defines it in his famous textbook *Fundamentals of Ecology*, "...[a]ny unit that includes all of the organisms (i.e. the 'community') in a given area interacting with the physical environment, so that a flow of energy leads to clearly defined trophic structure, biotic diversity, and material cycles (i.e. exchange of materials between living and nonliving parts) within the system is an ecosystem."[50] H. T. Odum extends this thinking further to argue that "energy, ecology, and economics form a single, unified system."[51] Such a perspective on systems also introduces what philosophers of Object-Oriented Ontology (OOO) refer to as a "flat ontology," a model that places all objects on the same agential playing field, challenging hierarchical systems.[52] For example, in *Intensive Science and Virtual Philosophy* Manuel DeLanda explains that "while an ontology based on relations between general types and particular instances is hierarchical, each level representing a different ontological category (organism, species, genera), an approach in terms of interacting parts and emergent wholes leads to a flat ontology, one made exclusively of unique, singular individuals, differing in spatio-temporal scale but not in ontological status."[53] This level footing can exacerbate the problems of spatiotemporal scale that are taken up in Chaps. 4 and 5. For now, it is enough to say that the flat ontology of OOO can tend to oversimplify relationality and agency, which Latour responds to with his conception of ANT.

These tensions are similarly present in the technocratic rhetoric of ecosystems. As Taylor explains, "...[t]echnocrats believe they can handle social complexity in a value-free manner, maintaining a distance from specific interests and political details."[54] The seamlessness that technocrats believe that solutionism will bring about ignores the messy complexities that spatial and temporal scales introduce to ecological thinking. H. T. Odum explains that Silver Springs is an ideal site for analysis because the literature suggests that the site has not changed since the 1860s.[55] Yet, this "steady state" locks the place into the kind of spatiotemporal regimes that I discuss later, where time is thought to stand still, or is artificially removed from the equation, and place becomes a symbolic abstraction. At the same time, Odum notes the potential for future disruption of this state by the growth of industry and development, pointing out other Florida springs which have ceased flow.[56] Yet, while the energy rhetoric that H. T. and Eugene Odum espoused throughout their careers—building from Lindeman's ecosystem concept—seems to reject the organismal approaches of earlier ecological theories, the influence of Fredrick Clements' organicism clearly reflects many of the holistic and teleological perspectives. Dana Phillips argues that "Odum's descriptions of ecology as a discipline have a figurative dimension and a Clementsian flavor at odds with his professed allegiance to the ecosystem concept."[57] While Tansley's "ecosystem" replaced organicist holism with a materialist ontology, its deployment by the Odum brothers carried forward remnants of earlier ecological theories. Reflecting on this conceptual contortion, Phillips remarks that "[Eugene] Odum's work demonstrates how stubbornly persistent analogies can be."[58] Energy rhetoric served to elide the connections of "new ecology" to the older concepts of the discipline. At the same time, this energy rhetoric served to root ecosystems ecology in the atomic age. Eugene Odum writes that "The new ecology is rooted in a solid historical development, but its rise to a front-line position in man's thinking is a consequence of the exploitation of atomic energy."[59] As such, the new ecology is directly connected to nuclear technology and energy at its core.

Along similar lines to ecosystems ecology, contemporary rhetorical scholarship is likewise haunted by the capacious and fraught modern conceptions of energy. Ingraham traces the interwoven threads of energy and rhetoric to Kennedy's "Hoot," which argues, among other things, that rhetoric might best be understood as a form of energy exchange, that rhetoric's energy is "perhaps a special case of the energy of all physics."[60] As a decorated scholar of Aristotelian rhetoric, Kennedy was steeped in his energetic theories, and the relationship between potentiality and actuality deeply informs Kennedy's understanding of rhetorical *transmission*, which produces a definition that includes nonhumans as capable rhetoricians. He specifically discusses the "complex code of signs" used by animals, giving examples of birds and primates, but he also does not close off this definition to plants, fungi, and bacteria. In broadening the boundaries of rhetoric beyond the human, Kennedy draws from evolutionary and social biology, linguistics, grammatology, and physics. While his focus is on living organisms, he does not discuss how viruses might complicate or figure into his definition. Importantly, Kennedy suggests the "rheme" as a "unit of rhetorical energy."[61] Mapping Kennedy's rheme onto Odum's emergy brings the energetic connections between ecosystems ecology and rhetorical ecologies into clearer perspective. If emergy allowed for a holistic view of ecological systems—one that considers not only the immediate biological organisms in a specific place, but also the broad transformations of energy exchanges that sustain them—so might the rheme allow rhetoric scholars to expand how we imagine and study rhetorical circulation and transformation within networks.

Ingraham connects Kennedy's work with a paradigm shift in rhetoric, examining the ways that rhetorical ecologies,[62] circulation,[63] and even "kinetic energy"[64] disrupt prior notions of the "rhetorical situation."[65] Essentially, the basic critique of the "situation" metaphor holds that it is too static, simplistic, abstract, and places too much emphasis on the agency of the speaker/author of the discourse, when much of the persuasive capacity of a message is not so simply within a speaker's control. Specifically, transformation leads to a distributed model of rhetoric that "does not derive solely from an actualized image's rhetorical design, nor is it a static affair, especially when it comes to [...] viral circulation."[66] As Ingraham notes, for recent scholars,[67] energy has become an "organizing concept" in the related concepts of circulation studies and rhetorical ecologies.[68] Alongside ecology, energy helps rhetoricians characterize discourse through more nuanced means, allowing rhetorical inquiry to better account for the complexity of things like digital networks.

While the shift toward studying ecologies, circulation, and complex systems has undoubtedly yielded a rich subfield within rhetoric—one that continues to grow and thrive—there remains a need to address the problems and concerns that ecology and energy import along with the metaphor. For example, Justine Wells examines how Kennedy's definition of rhetoric as a "mechanism for survival," and its function as promoting "the survival of the fittest," engages with Social Darwinism that centers "whiteness and white supremacy."[69] By investigating the genealogy of the ecosystem, this book seeks to decenter such perspectives though a conceptual history with the ultimate goal of bettering ecological and rhetorical inquiry. As such, energy—and specifically the node I identify between the rheme and emergy—indexes one important place to locate overlapping problems and approaches that can help map rhetorical ecologies onto ecosystems ecology, connections and troubles that I take up in the next chapter and develop throughout the remainder of the book. As such, energy rhetoric serves to forward the technomorphic and technocratic rhetoric of ecosystems.

Ecosystem as Diagram and Apparatus

Conceptually, the ecosystem functions less like a metaphor for complexity, circulation, and networks and more like a diagram, as defined by Deleuze and Guattari. In their definition, the diagram's function is greater than a representation or scaled model. Rather, diagrams are systems of meaning-making. Deleuze discusses the diagram in *Difference and Repetition* as "a system of multiple, non-localizable connections between differential elements which is incarnated in real relations and actual terms."[70] Diagrams actively participate in the processes, practices, and production of meaning. Byron Hawk and Matthew Halm have each investigated the diagram as a key term for rhetoric, situating diagrammatic rhetoric as a means for resisting treating concepts as metaphors or abstract models. Halm gives the example of plate tectonics, where the metaphor of an *earthquake as communication* is contrasted with the ways that the "diagram of the mechanism that drives the circulation of tectonic plates can produce an understanding of the mechanism that drives the circulation of rhetoric."[71] In other words, diagrams not only act as representations, but also rather actively function to generate meaning. Hawk gives the example of virality, which "isn't just a metaphor for how discourse circulates" but rather functions as "a model for how forces circulate through all kinds of encounters."[72] As a diagram, the ecosystem concept participates in the process of making meaning—just as environment, ecology, and *Umgebung* each generate new ways of thinking.

Throughout their career, the Odum brothers produced numerous figures and illustrations representing a wide array of abstruse ecological concepts. Throughout their body of work, they demonstrate a keen understanding of visual rhetoric and design, and in correspondence with publishers, Eugene Odum expressed a strong interest in how visuals can communicate science to non-specialist audiences. As Phillips puts it, "Odum's illustrations are best regarded as mnemonic devices and pedagogical aids, and not as 'realistic' depictions of the natural world" and clearly "are a poor sort of visual shorthand with which to convey some extremely recondite ideas."[73] As such, he argues that "[e]cosystem modeling seems to be essentially rhetorical, in that the persuasive power of model ecosystems tends to be more important than the accuracy of their details."[74] Building from his work, this book argues that ecosystems function as a diagram that has been extremely influential, not only in shaping perceptions of relationships between organisms and environments in the hard sciences, but also in defining and directing inquiry in the humanities and social sciences. Yet, such an application further tangles the ecosystem with communications and rhetoric scholarship, as the ecosystem concept is deeply rooted in both cybernetics and information theory, born out of work in communications during WWII.[75] In many ways, the connections between cybernetics, communication, ecology, and the ecosystem are stacked around contemporary Rhetoric, Writing, and Communication Studies (RWCS) research like nesting dolls, each inside of the other, where communication networks serve as a diagram for ecology, producing the ecosystem, which in turn functions as a diagram for how we imagine digital rhetoric and networks as we play Ring Around the Rosie.

For example, Laurie Gries and Collin Gifford Brooke's 2018 edited volume *Circulation, Writing, and Rhetoric* positions *circulation* as an emerging threshold concept for RWCS scholarship, specifically discussing the relationship between ecological rhetorical models and the concepts of circulation, flow, distribution, complexity, and transformation.[76] These conceptions of ecology within the growing research area of circulation studies can be understood using the ecosystem as a diagram for ecological rhetorical inquiry. Approaching rhetoric through circulation moves beyond drawing metaphoric parallels between ecology and rhetoric. Instead, the ecosystem becomes a functional diagram that actively shapes how DH scholars approach and study rhetorical circulation, such as in metaphors of trends and virality.[77] By understanding the ecosystem—with its complex history and rich body of knowledge—as a diagram, researchers can approach the dynamic interactions and flows of information within rhetorical ecologies in a more material and process-oriented manner. To further complicate the tangled diagrammatic rhetoric that connects information theory, ecosystems, and communication, the concept is also embedded within and endued with the characterization of the ecosystem as a cybernetic apparatus and machine, a topic richly covered in Thomas Pringle, Gertrud Koch, and Bernard Stiegler's *Machine*. Building from Guattari, Pringle focuses on the influence of cybernetics on environmental politics, specifically as it shaped a focus on environmental control and resource management, in which the ecosystem "subsequently becomes the dominant concept for describing biophysical reality as a cybernetic hybrid of nature and machine, otherwise, as an amalgamation of technological *and* ecological systems."[78] Importantly, Pringle invokes Bernard Geoghegan's work with the cybernetic apparatus[79] to understand the shift in scientific practices that took place after WWII. Apparatus identifies the ecosystem as more than just a metaphor. Rather, it produces the tensions Pringle identifies between "ecology and economy" that are "later repressed and packaged for use by cybernetic universalism.[80] Ecosystems present a coherent conceptual model, but as an apparatus it also obscures the sometimes conflicting elements of technology, ecology, and communication systems.

In this sense, the ecosystem as apparatus refers not only to its metaphors of cybernetic systems, its focus on instrumentation, or to procedures from mathematics and engineering, but also to the diagrammatic rhetoric that allowed these components to function together at the conceptual level. By viewing the ecosystem as an apparatus, the diagrammatic rhetoric that allows it to transcend its ecological context and integrate cybernetics, economics, and communications comes into view within the context of the Pacific theater of WWII, a topic I will return to in the next chapter. As Pringle makes clear, the blending of what Hörl refers to as "the nature / technics divide" challenged traditional boundaries between the natural world and the machine, rhetorically reshaping how we understand Guattari's "three ecologies" (social, mental, and environmental).[81] The diagrammatic rhetoric of the ecosystem apparatus becomes a site where scientific, technological, and social processes converge, emphasizing the entangled role that material and symbolic elements like energy play in the formation of modern ecological inquiry. Ultimately, understanding how the diagrammatic rhetoric of the ecosystem apparatus serves as a framework for management and control both enriches our understanding of the historiography of science and serves as a means of understanding how the ecosystem concept persists in shaping how we conceptualize and realize digital and ecological methods in RWCS and beyond.

Energy and Consciousness

In the spring of 1975, a group of beat poets including Allen Ginsberg, Michael McClure, and Gary Snyder came together with H. T. Odum for a week-long celebration of "Energy and Consciousness," jointly hosted by the English Department and the Engineering School at the University of Florida (Fig. 2.3). The events featured talks on energy and poetry by Ginsberg, McClure, and Snyder, as well as a panel on energy and consciousness with Odum, Ginsberg, McClure, and the nuclear physicist Henry Gomberg. Odum saw the event as an interdisciplinary "experiment in bridging the gap between science and the arts."[82] Featured among the

Fig. 2.3 Ad promoting the 1975 Energy and Consciousness event at the University of Florida that featured H. T. Odum and beat poets (University of Florida 1975)

events that took place that week was a tour of an experiment to test the concept of "ecosystem services" that Odum had been conducting in a nearby wetland involving dumping raw municipal waste into a nearby cypress dome in order to test the effectiveness and efficiency of nature-based waste processing, compared to traditional waste treatment practices.[83]

This event demonstrates deep connections between ecosystems and the Beatnik movement as part of the development of radical environmentalism in the 1970s. Poets like Ginsberg and Snyder drew upon rhetorical energy in concert with ecosystems in their verse to promote environmental consciousness. At the same time, energy serves to flatten environmentalism by reducing relationality through the rhetoric of systems. Odum's research at Silver Springs and in wetlands near Gainesville has made important contributions to environmental research and culture. Further, these sites have long participated in the diagrammatic rhetoric through which we construct the so-called nonhuman world. Settler colonial removal of Indigenous peoples, including the Seminole Tribe, contributed to an image of the natural world devoid of human presence, contributing to the Romantic views of the natural world and the frontier wilderness myth that I discuss in the next chapter. Following the developments of communications and nuclear technology in the aftermath of WWII, the natural world was transformed into a cybernetic machine through the ecosystem concept, with the presumption that its electrical flows could be managed and controlled. Odum's research at Silver Springs was a pivotal moment in ecosystems ecology. His groundbreaking work helped conceptualize ecosystems as dynamic, interconnected systems of energy exchange.

Energy flow, both for ecology and for rhetoric, offers capacious and sometimes fraught metaphors for complex systems. In *Energy Dreams*, Marder explains that the concept of energy calls for us to develop a "nonviolent framework for thinking about and practically relating to energy without destroying living beings and our planet through its extraction."[84] In the period of time known as the Anthropocene, technological and anthropogenic changes have drastically altered environments in a relatively short period of time. Today, the Silver Springs study offers a baseline for understanding how large-scale changes are affecting local environments. Odum's use of energy as a metaphor for ecosystem dynamics demonstrates deep connections between ecology and rhetoric, revealing a need for interdisciplinary approaches to address the most urgent environmental and communication challenges. This example reminds us that apprehending and addressing systemic problems requires a richer understanding of scientific invention through the perspectives of RWCS. In the following chapters, I build from this study to examine the convergence of ecology, technology, and communication as they present challenges and opportunities for environmental communication and ecological inquiry.

Notes

1. William Bartram, *Travels Through North & South Carolina, Georgia, East & West Florida, the Cherokee Country, the Extensive Territories of the Muscogulges, or Creek Confederacy, and the Country of the Chactaws; Containing An Account of the Soil and Natural Productions of Those Regions, Together with Observations on the Manners of the Indians* (James & Johnson, 1791), 167.
2. Ibid., 168.
3. Judith Magee, *The Art and Science of William Bartram* (Penn State University Press, 2007).
4. Matthew Wynn Sivils, "William Bartram's *Travels* and the Rhetoric of Ecological Communities," *ISLE: Interdisciplinary Studies in Literature and Environment* 11, no. 1 (2004): 57–70, doi:10.1093/isle/11.1.57
5. See Lu Vickers and Cynthia Wilson-Graham, *Remembering Paradise Park: Tourism and Segregation at Silver Springs* (University Press of Florida, 2015). Silver Springs Park was segregated, leading to the creation of Paradise Park in 1949, which split the area around the spring until 1967. Vickers and Wilson-Graham connect the iconic location to the work of novelist and anthropologist Zora Neale Hurston and other important African American writers. Their book traces the rhetorical lives of the springs, from the segregated park to the water that circulated independently of these boundaries. See Zora Neale Hurston, *Seraph on the Suwanee* (Harper Perennial, 1900), 294–295. Hurston draws on the underground flow of water in *Seraph on the Sewanee*, which articulates the experience of "Florida crackers" in a time of segregation. In the novel, her character Jim claims that "Some folks are surface water and are easily seen and known about. Others get caught underground, and have to cut and gnaw their way out if they ever get seen by human eyes." Silver Springs was but one site where racism held many African Americans down in unseen depths.
6. Wendy Adams King, "Through the Looking Glass of Silver Springs: Tourism and the Politics of Vision," *Americana: The Journal of American Popular Culture (1900–present)* 3, no. 1 (2004): para. 3, https://americanpopularculture.com/journal/articles/spring_2004/king.htm
7. Ibid.
8. Arthur Tansley, "What Is Ecology?," *Biological Journal of the Linnean Society* 32 (1987): 11. This paper was originally published in 1951 as a pamphlet by the Council for the Promotion of Field Studies, which became the Field Studies Council.
9. For a detailed study of the racist origins of natural history, as well as the contemporary problems facing ecological inquiry, see Maria N. Miriti, Ariel J. Rawson, and Becky Mansfield, "The History of Natural History

and Race: Decolonizing Human Dimensions of Ecology," *Ecological Applications* 33, no. 1 (2023): e2748, doi:10.1002/eap.2748
10. For example, see Donald Worster, *Nature's Economy: A History of Ecological Ideas* (Cambridge University Press, 1994).
11. Peder Anker, *Imperial Ecology Environmental Order in the British Empire, 1895–1945* (Harvard University Press, 2001): 4.
12. For more on Bartram's influence on ecological inquiry, see Matthew Wynn Sivils, "William Bartram's *Travels* and the Rhetoric of Ecological Communities," *ISLE: Interdisciplinary Studies in Literature and Environment* 11, no. 1 (2004): 57–70, doi:10.1093/isle/11.1.57
13. See Peter Ayres, *Shaping Ecology: The Life of Arthur Tansley* (Wiley-Blackwell, 2012): 15–19.
14. Ibid., 16.
15. For more on Haeckel and the invention of the term "ecology," see Robert C. Stauffer, "Haeckel, Darwin, and Ecology," *The Quarterly Review of Biology* 32, no. 2 (1957): 138–144, doi:10.1086/401754
16. See W. M. Kemp and W. R. Boynton, "Productivity, Trophic Structure, and Energy Flow in the Steady-state Ecosystems of Silver Springs, Florida," *Ecological Modelling* 178, no. 1–2 (2004): 43–49, doi:10.1016/j.ecolmodel.2003.12.020
17. H. T. Odum, "Trophic Structure and Productivity of Silver Springs, Florida," *Ecological Monographs* 27 (1957): 55, doi:10.2307/1948571
18. Robert L. Knight, "Energy Basis of Control in Aquatic Ecosystems" (PhD diss. University of Florida, 1980).
19. D. A. Munch et al., *Fifty-year retrospective study of the ecology of Silver Springs, Florida.* St. Johns River Water Management District (Special Publication SJ2007-SP4), 2006.
20. Caroline Gottschalk Druschke, Laura A. Meyerson, and Kristen C. Hychka, "From Restoration to Adaptation: The Changing Discourse of Invasive Species Management in Coastal New England Under Global Environmental Change," *Biological Invasions* 18 (2016): 2740, doi:10.1007/s10530-016-1112-7
21. Donnie Johnson Sackey, *Trespassing Natures: Species Migration and the Right to Space* (Ohio State University Press, 2024), 3.
22. Ibid., 4.
23. Peter J. Taylor, *Unruly Complexity Ecology, Interpretation, Engagement* (The University of Chicago Press, 2005), 52.
24. Ibid., 74.
25. Ibid., 60.
26. Ibid., 61.
27. H. T. Odum, "Emergy in Ecosystems," in *Environmental Monographs and Symposia*, ed. N. Polunin (John Wiley, 1986), 337–369.

28. Madison Jones, "The Energy of Place in Florida Springs Activism," in *Grassroots Activisms: Public Rhetorics in Localized Contexts*, Lisa L. Phillips, Sarah Warren-Riley, and Julie Collins Bates, eds. (The Ohio State University Press, 2024), 128–144.
29. Ibid., 133.
30. Chris Ingraham, "Energy: Rhetoric's Vitality," *Rhetoric Society Quarterly* 48, no. 3 (2018): 260–268, doi:10.1080/02773945.2018.1454188
31. Michael Marder, *Energy Dreams: Of Actuality* (Columbia University Press, 2017), 2.
32. Ibid., 7.
33. Ingraham, "Energy: Rhetoric's," 262.
34. Odum, "Trophic Structure."
35. Ibid., 56.
36. H. T. Odum, "Self-Organization and Maximum Empower," in *Maximum Power: The Ideas and Applications of H.T. Odum*, ed. Charles A. S. Hall (University Press of Colorado, 1995).
37. Patrick Kangas, "The Role of Passive Electrical Analogs in H. T. Odum's Systems Thinking," *Ecological Modelling* 178, no. 1–2 (2004): 101, doi:10.1016/j.ecolmodel.2003.12.019
38. Ibid.
39. Sidney I. Dobrin and Madison Jones, eds., *Rhetorical Ecologies* (National Council of Teachers of English Press, 2024); Enrich Hörl, *General Ecology: The New Ecological Paradigm* (Bloomsbury, 2017).
40. George Kennedy, "A Hoot in the Dark: The Evolution of General Rhetoric," *Philosophy and Rhetoric* 25, no. 1 (1992): 1–21, https://www.jstor.org/stable/40238276
41. B. Å. Månsson and J. M. McGlade, "Ecology, Thermodynamics and H.T. Odum's Conjectures," *Oecologia* 93 (1993): 582–596, doi:10.1007/BF00328969
42. Ibid., 589.
43. Debra Hawhee, "Kairotic Encounters," in *Perspectives on Rhetorical Invention*, Janet M. Atwill and Janice M. Lauer, eds. (University of Tennessee Press, 2002), 16–35; Ehren Helmut Pflugfelder, *Geoengineering, Persuasion, and the Climate Crisis: A Geologic Rhetoric* (University of Alabama Press, 2022); Thomas Rickert, *Ambient Rhetoric: The Attunements of Rhetorical Being* (University of Pittsburgh Press, 2013).
44. Odum, "Self-Organization," 311.
45. Ibid.
46. David M. Scienceman, "Letters to the Editor: Emergy Definition," *Ecological Engineering* 9 (1997): 209–212; Odum, "Emergy in Ecosystems."

47. J. Cadzow, "Dr Scienceman's brave new word," *The Australian* (May 15, 1984): 7.
48. Scienceman, "Letters to," 209.
49. H. T. Odum, "Energy, ecology and economics," *AMBIO* 2, no. 6 (1973): 224.
50. Eugene P. Odum, *Fundamentals of Ecology* (W. B. Saunders, 1953), 8.
51. H. T. Odum, "Energy, ecology," 220.
52. See Levi R.Bryant, *The Democracy of Objects* (Open Humanities Press, 2011).
53. Manuel DeLanda, *Intensive Science and Virtual Philosophy* (Bloomsbury, 2002), 47.
54. Taylor, *Unruly Complexity*, 53.
55. Odum, "Trophic Structure," 57.
56. Ibid., 58.
57. Dana Phillips, *The Truth of Ecology* (Oxford University Press, 2003), 64.
58. Ibid.
59. Eugene Odum, "The New Ecology," *BioScience* 14, no. 7 (1964): 14, https://doi.org/10.2307/1293228
60. Kennedy, "A Hoot," 13.
61. Ibid., 2.
62. Jenny Edbauer, "Unframing Models of Public Distribution: From Rhetorical Situation to Rhetorical Ecologies," *Rhetoric Society Quarterly* 35, no. 4 (2005): 5–24, doi:10.1080/02773940509391320
63. Laurie Gries, *Still Life With Rhetoric: A New Materialist Approach for Visual Rhetorics* (Utah State University Press, 2015).
64. Carolyn R. Miller, "What Can Automation Tell Us About Agency?," *Rhetoric Society Quarterly* 37, no. 2 (2007): 147, https://doi.org/10.1080/02773940601021197
65. For example, see Lloyd F. Bitzer, "The Rhetorical Situation," *Philosophy & Rhetoric* 1, no. 1 (1968): 1–14, mailto:2@638457_1_En.docx; Richard E. Vatz, "The Myth of the Rhetorical Situation," *Philosophy & Rhetoric* 6, no. 3 (1973): 154–61,http://www.jstor.org/stable/40236848
66. Gries, *Still Life*, 27.
67. For example, Catherine Chaput, "Rhetorical Circulation in Late Capitalism: Neoliberalism and the Overdetermination of Affective Energy," *Philosophy & Rhetoric* 43, no. 1 (2010): 1–25, doi:10.5325/philrhet.43.1.0001
68. Ingraham, "Energy: Rhetoric's," 263.
69. Justine Wells, "The Energy of Whiteness," Presentation at 20th Biennial Conference of the Rhetoric Society of America, Baltimore, MD, May 2022, para. 3,https://rhetoricsociety.confex.com/rhetoricsociety/2022/meetingapp.cgi/Session/1616

70. Gilles Deleuze, *Difference and Repetition*, trans. P. Patton (Columbia University Press, 1994), 183.
71. Matthew Halm, "Molten Circulation and Rhetoric's Materiality," *Enculturation* 35 (2023): para. 5, https://enculturation.net/molten_circulation
72. Byron Hawk, "A Diagrammatics of Persuasion," in *Circulation, Writing, and Rhetoric*, Laurie Gries and Collin Gifford Brooke, eds. (Utah State University Press, 2018), 313.
73. Phillips, *The Truth*, 63.
74. Ibid., 69.
75. Robert E. Ulanowicz, "Information Theory in Ecology," *Computers & Chemistry* 25, no. 4 (2001): 393–399, doi:10.1016/S0097-8485(01)00073-0
76. Laurie Gries and C. G. Brooke, eds., *Circulation, Writing, and Rhetoric* (Utah State University Press, 2018).
77. Nicholas Van Horn, Aaron Beveridge, and Sean Morey, "Attention Ecology: Trend Circulation and the Virality Threshold," *Digital Humanities Quarterly* 10, no. 4 (2016), https://www.digitalhumanities.org/dhq/vol/10/4/000271/000271.html
78. Thomas Pringle, Gertrud Koch, and Bernard Stiegler, *Machine* (Meson Press; University of Minnesota Press, 2019), 54.
79. For a robust discussion of the apparatus, based on Michel Foucault's conception of the term, which engages with its development out of information theory and communication, see Bernard Dionysius Geoghegan, "From Information Theory to French Theory: Jakobson, Lévi-Strauss, and the Cybernetic Apparatus," *Critical Inquiry* 38 (2011), doi:10.1086/661645. He explains that apparatus identifies more than a discrete metaphor or technology. Rather, it names systems among numerous and diverse relational elements, from institutions to ideology to instruments.
80. Pringle, Koch, and Stiegler, *Machine*, 66.
81. Hörl, *General Ecology*; Félix Guattari, *The Three Ecologies* (Continuum, 2008).
82. University of Florida, "Science, Poetry Linked in Focus on Energy Woes," *The Independent Florida Alligator*, April 18, 1975, 13, https://ufdcimages.uflib.ufl.edu/UF/00/02/82/90/00743/00743.pdf
83. For a firsthand account and critical reflection on this event, see Robert Walker, "Energy and Consciousness," *Beatdom*, May 2016, 17, https://www.researchgate.net/publication/303382212_Energy_and_Consciousness
84. Marder, *Energy Dreams*, i.

Open Access This chapter is licensed under the terms of the Creative Commons Attribution-NonCommercial-NoDerivatives 4.0 International License (http://creativecommons.org/licenses/by-nc-nd/4.0/), which permits any noncommercial use, sharing, distribution and reproduction in any medium or format, as long as you give appropriate credit to the original author(s) and the source, provide a link to the Creative Commons license and indicate if you modified the licensed material. You do not have permission under this license to share adapted material derived from this chapter or parts of it.

The images or other third party material in this chapter are included in the chapter's Creative Commons license, unless indicated otherwise in a credit line to the material. If material is not included in the chapter's Creative Commons license and your intended use is not permitted by statutory regulation or exceeds the permitted use, you will need to obtain permission directly from the copyright holder.

CHAPTER 3

The Trouble with Ecology: The Afterlives of Radioecology in the Marshall Islands and Puerto Rico

Abstract Focusing on the colonial and military context of ecosystems ecology, this chapter examines radioecology alongside nuclear testing in the Marshall Islands and Puerto Rico. The US Atomic Energy Commission funded ecological research to understand how radioactive pollution impacts environments and people. During this time, the "new ecology" became enfolded in the politics of militarized nuclear colonialism. The legacy of radioecology research persists in influencing ecological science today, shaping narratives of environmental management, resilience, disturbance and restoration ecology, and risk communication.

Keywords Historiography • Nuclear colonialism • Radioecology • Marshall Islands • Puerto Rico • Environmental justice

[T]he trouble with *wilderness* is that it quietly expresses and reproduces the very values its devotees seek to reject. The flight from history that is very nearly the core of wilderness represents the false hope of an escape from responsibility, the illusion that we can somehow wipe clean the slate of our past and return to the tabula rasa that supposedly existed before we began to leave our marks on the world. —William Cronon, "The Trouble with Wilderness"[1]

© The Author(s) 2026
M. P. Jones, *Inventing Ecosystems*, Palgrave Studies in Media and Environmental Communication,
https://doi.org/10.1007/978-3-031-98793-9_3

El Verde Field Station, Río Grande, Puerto Rico, United States
The road to El Yunque snakes and climbs as I make my way from the outskirts of Loíza. As we wind our way into Bosque Nacional, vegetation along the roadway grows more dense. I roll the window down to let in the humid scent of earth and blossom. It has been nearly ten years since I was last here. As those memories come pouring back, so does the realization of just how much has changed in that time, both in my life and in the place itself. Looking back, things felt so much lighter as my wife and I wandered the dark, wet trails lined with towering tabonuco trees (Dacryodes excelsa), without a thought to how our kids were doing with their grandparents or if there were enough snacks and toys to occupy them for the day. I remember thinking that the forest felt timeless, standing like an eternal picture of itself, as the steady rhythms of rain met with the patient growth of ancient flora. Now, I realize that such a view of the forest marked me as an outsider, one who wasn't aware of the profound ways the forest has changed over different timescales.[2] Today, I see the marks of time written differently in the landscape. Just a few months after my last visit, Hurricane Maria tore through the forest on September 20, 2017, uprooting many of the old-growth hardwoods and cutting the tops of nearly all of the tree canopy, exposing the forest floor to increased light that allowed grasses and shrubs to grow. The storm claimed nearly 3000 human lives in Puerto Rico, leveled buildings, flooded roads, obstructed access to drinking water, and took out the power grid on parts of the island for nearly a year. Comparative studies have found that Maria (category 4) was far more ecologically destructive than other large hurricanes that made landfall in the recent past, such as Hugo (category 3) in 1989 and Georges (category 3) in 1998.[3] I recall the footage of the devastation I had seen and brace myself for what I will witness. Yet, as we make our way into the Luquillo Experimental Forest, I am relieved to see how much has recovered from the devastating damage in less than a decade.

  It occurs to me that this transformation is a striking example of what ecologists call "disturbance." Disturbance ecology has emerged as a subfield that grew in part from the research on ecosystems that H.T. Odum led, along with a massive team of researchers at the El Verde Research Station, where I will stop on my trip today. This research station is part of the NSF's Long-Term Ecological Research (LTER), a network of ecological research sites that focus on long-term and large-scale phenomena. Odum and his colleagues have spent decades researching how forests recover from natural disasters, from the biochemical scale to the level of ecological communities. Yet, disturbance in El Yunque has not come only from natural causes. During the Cold

War, El Yunque was a key site for Odum's radioecology research beginning in the 1940s (though Odum would continue to do research there for the rest of his life) and developing in the 1950s and 60s with support from the Rockefeller Foundation and approximately one million dollars in grants from the AEC over a four-year period.[4] *In 1964, Odum's team introduced a cylinder containing 10,000 curies of cesium chloride powder as a gamma radiation source, which was deposited in the forest by helicopter. Armed guards with dogs patrolled a fence line with bilingual signs to ensure that no people would enter the radiation zone. As I hike up to the site, I notice a piece of the old fence, including a metal tag that might bear Odum's own handwriting. This experiment brought major insights into how an ecosystem responds to stress and even disaster, as well as how radiation is dispersed in the environment, but this project reflects deeper tensions between scientific research, military interests, and the local community. Many Puerto Ricans viewed these experiments as an extension of occupation by the US military, part of a long history of Puerto Ricans being subjected to unethical experimentation by both the US government and medical researchers.*[5] *Today, El Verde remains a leading research site for ecologists to understand how tropical ecosystems respond to ecological disturbances in order to predict how they might respond to the increasing effects of climate change in the future. However, the uneasy legacy of US exploitation also shapes public perception of science in places like Puerto Rico. This legacy, coupled with the emerging long-term impacts of climate change—such as the devastation wrought by Maria—demonstrates the urgent need for ecological inquiry to better engage with environmental justice.*

The Trouble with Ecology

In his pivotal essay, "The Trouble with Wilderness; or Getting Back to the Wrong Nature," environmental historian William Cronon interrogates the concept of wilderness as a cultural construct that continues to shape environmentalist thinking. Through history, Cronon is able to situate the problems that nature presents as a signifier for US cultural values and environmental ethics. He demonstrates how pristine nature functioned as a concept that supported frontier colonialism, erasing Indigenous peoples by supporting the myth of unpopulated wilderness. His analysis was part of a paradigm shift for first-wave ecocriticism, moving away from accepting the construction of nature as a Burkean god term in nature writing and toward more careful engagement with the complex ways that language structures relations with the natural world and/or the nonhuman.

In this section, I share with Cronon what he calls "common ground," or what rhetoricians call a "commonplace." Were we to replace "wilderness" with "ecology" in the epigraph for this chapter, we would start to get at the trouble with modern ecology. As Cronon puts it, "…the modern environmental movement is itself a grandchild of romanticism and post-frontier ideology."[6] As such, ecology bears what Ludwig Wittgenstein calls "family resemblances"[7] to the wilderness concept, resulting in a reproduction of some of the same conceptual and material problems that Cronon demonstrates with colonial constructions of "Nature." Ecosystems serve similarly to "tame" the natural world[8] even as the concept can serve to remove "native peoples […] from an ancient home" by adopting managerial functions for pristine environments.[9] Because the Odums' research required sites that were relatively stable, and to some extent undisturbed by human activity, as well as isolated and remote, the ecosystem required a very particular type of environment, and as Cronon notes, research at sites like the tropical rainforest contributed to what would become "since the 1970s […] the most powerful icon of unfallen, sacred land—a veritable Garden of Eden—for many Americans and Europeans," while contributing at the same time to the view that "protecting the rain forest in the eyes of First World environmentalists all too often means protecting it from the people who live there."[10]

The trouble that Cronon identifies can be observed in the Odums' radioecology experiments in the coral reefs at Enewetak in the Marshall Islands and as part of the Rain Forest Project[11] in the Luquillo Experimental Forest in El Yunque National Forest, Puerto Rico (Fig. 3.1). At these sites, large-scale ecological research was led by predominantly white American researchers who sought to understand how ecosystems functioned in order to better manage and control them, while largely ignoring the lived experiences of the local inhabitants.[12]

By securing major funding for the research, seizing the opportunity to further test and develop theories of energy dynamics, developing methods for replicable radioecology studies, and revolutionizing ecological fieldwork, the Rain Forest Project made major contributions to the paradigm shift of ecosystems ecology.[13] Reflecting on the work they had conducted at Enewetak, Eugene Odum refers to the coral reef ecosystem "as a kind of oasis in a desert" that "can stand as an object lesson for man who must now learn that mutualism […] between producers and consumers, coupled with efficient recycling of materials and use of energy, are the keys to maintaining prosperity in a world of limited resources."[14] As such, this

Fig. 3.1 "Source and Crew. El Verde. June 1965." The crew at El Verde, led by Howard T. Odum, with a cylinder that contains a 10,000-curie cesium gamma radiation source. The source was deposited in the forest by helicopter in 1964 as part of a radioecology study that took place from 1962 until 1970 and was funded by the AEC. Courtesy of the Howard T. Odum Papers, Special and Area Studies Collections, George A. Smathers Libraries, University of Florida, Gainesville, Florida

pristine and supposedly isolated environment allowed the Odums to understand how energy moves through a closed system. Yet, it was precisely human impacts on the environment that led the Atomic Energy Commission (AEC) to recruit H. T. Odum to research the impact that testing nuclear bombs had on the environment (Fig. 3.2).

The tensions between ecosystems research, militarized colonialism, and technocratic managerialism become particularly stark when examining the context of nuclear testing sites. Aimee Bahng illustrates these tensions in her discussion of the Marshall Islands as both ecological laboratory and site of nuclear imperialism. Building from Laura J. Martin, Bahng explains that "the uncomfortable convergence of nuclear proliferation and environmental conservation was not correlational but perhaps even causal, bound together by the impetus to promise containment of ecological fallout. Securitization thus becomes the rationale for both the destruction

Fig. 3.2 An aerial photograph of Runit Dome (sometimes referred to as Cactus crater containment structure) on Runit Island, Enewetak Atoll. The crater that was created by the United States during the detonations of Operation Hardtack I in 1958 was used to create a burial pit to cover the radioactive material that was scraped from the contaminated Enewetak Atoll islands. Courtesy of Wikimedia Commons, https://commons.wikimedia.org/wiki/File:Runit_Dome_001.jpg (Public Domain)

and the repair, for both debilitation and the demand to overcome."[15] At the proving grounds in the Marshall Islands, the Odum brothers' research became a site where complex dynamics converged—where ecological theory, environmental management and conservation, and nuclear imperialism became enfolded. At the same time, the Odums were committed to using funding not only to meet the aims of the AEC, but also to test their theories of ecosystems and other ecological processes in what Donald Worster refers to as an "ideally isolated laboratory."[16] As scientists who worked with H. T. in Puerto Rico recount, his approach to large-scale research projects was to split his efforts fifty-fifty between the aims of the funding body and his own interests in ecological research. His older brother Eugene, who collaborated on the Rain Forest Project, would also find similar support from the AEC for his Savannah River Ecology Laboratory when DuPont (at the request of the US federal government) built an atomic weapons plant at the Savannah River Site. As such, the seemingly Edenic sites where many of the Odums' studies took place were also sites of extreme disturbance, occurring in the larger contexts of localized radiation experiments that impacted environments on a global scale.

At the same time, ecosystems seemingly offer a counterweight to what Cronon calls the "wilderness premise" which views environments as "remote from humanity and untouched by our common past."[17] In fact, the ecosystem concept as taken up by Eugene and H. T. Odum offers important lines of inquiry that directly, and productively, contradict the harmful ideological frameworks that position humans as unnatural, or as existing outside of the natural world. By viewing environments as machines, with calculable inputs and outputs, ecosystems ecology introduced a flat ontology to the nature/culture divide that undermined their separation in prior conceptions of Romantic Nature and Frontier Wilderness (a topic I will return to and expand on in the next chapter). This flattening allowed ecosystems to position nature and culture on a level playing field by examining how energy flowed between and among specific systems. In fact, H. T. Odum's work with ecological engineering sought to join "the economy of society to the environment symbiotically by fitting technological design with ecological *self design*."[18] The premise of ecological engineering, as envisioned by H. T. Odum, was that ecosystems could be made to directly "interface" with technology. This move reverses the removal of humans from nature in the wilderness concept by producing what Timothy Morton refers to as "ecology without Nature" which "takes nature out of the equation" by examining how it is constructed.[19]

Yet, just as wilderness removes humans from nature, so do ecosystems serve to remove both humans and nature from the imagined technics of ecological systems. Megan Raby lays bare the ways that the Rain Forest Project yields new insights about land-based approaches to environmental history. As Raby puts it, "Odum's ecosystem view allowed him to accept a field site that was not pure wilderness."[20] As Morton puts it, the "ecosystem becomes an immersive, impersonal matrix. Unfortunately for ideas of an ecological politics that would liberate us from the modern state, this is the systems thinking adopted by the RAND corporation [...]. Systems theory is holism without the sticky wetness, a cybernetic version of the ecological imaginary."[21] Systems replace the messy processes that humans and environments embody with cybernetic and energetic metaphors to characterize nature and culture. In doing so, these systems anthropomorphize machines, bodies, and the natural world. As Dana Phillips writes, the idea "...[t]hat society might be reorganized in accord with ecological principles was in fact a possibility that [Eugene] Odum, like most environmentalists, was eager to entertain."[22] Phillips offers a compelling analysis of how his metaphors, what I term "energy rhetoric" in the previous chapter, obscure a pernicious Malthusian ethic that overtly advocates for population control. The shift in metaphors elides these troubling assumptions and underscores the need to critically examine ecological and systems thinking. While the Odum brothers' energy rhetoric invites us to imagine a utopia where social dimensions are organized through ecological models and technocratic principles, it introduces troubling ethical concerns that reveal the limits of these frameworks. For ecological inquiry to move beyond these limitations, it is necessary to remap the conceptual history of ecologies and address the deeper rhetorical challenges posed by how we perceive and represent time and place within those ecologies.

By troubling ecology, I seek other ways to trace the conceptual history of rhetorical ecologies. In this section, I set out the twin rhetorical problems of ecology: the displacement of time through *kairos* and the derangement of place through scale. These two moves are foundational to ecosystems ecology—as well as systems thinking more broadly—and operate by reducing complexity through spatiotemporal slices. *Kairos* breaks from ordinary time while scale divides place into hierarchical dimensions. This phenomenon is what N. Katherine Hayles refers to as "making the cut,"[23] or later as "the Platonic backhand and forehand":

The Platonic backhand works by inferring from the world's noisy multiplicity a simplified abstraction. So far so good: this is what theorizing should do. The problem comes when the move circles around to constitute the abstraction as the originary form from which the world's multiplicity derives. Then complexity appears as a 'fuzzing up' of an essential reality rather than as a manifestation of the world's holistic nature. Whereas the Platonic backhand has a history dating back to the Greeks, the Platonic forehand is more recent.[24]

Along these lines, we may hear echoes of V. F. Cordova's critique of settler colonial metaphysics that produces a "strange definition [...]of man" in which "man is, at the same time, a pawn of the universe and its guardian."[25] Through temporal slice (*kairos*) and spatial slice (scale), ecosystems ecology studies the world as situated within these cuts, as if suspended between two pieces of glass in a microscope slide. As Dan Ehrenfeld argues, "...ecological models have emphasized change" but in doing so "they have deemphasized the distinct dynamics that animate the public sphere in particular times and places."[26] Within this frame, ecology relies on a particular space/time to reduce complexity in order to make holistic modeling possible. As a metaphor, rhetorical ecology inherits this methodology for organizing relations. These cuts allow ecologists to study systems, but they also displace and derange our sense of place and time. By decentering synchronic *kairos* along the same lines as Rachel Wolford's diachronic-synchronic model of agency,[27] the following two subsections demonstrate how these scalar problems (time and place) are both situated within the disciplinary history of ecosystems ecology and serve to displace the connections between ecology and nuclear colonial violence in rhetorical ecologies.

Motivated by a desire to prove that environments could be understood as systems, Eugene and H. T. Odum sought to demonstrate such a mechanical function in the natural world. Their approach reversed the typical purpose of fieldwork, where data collection leads to analysis and the development of theories. Instead, they assumed the ecosystem to be a *fait accompli*, a phenomenon that would naturally be instantiated given the proper conditions. Their search for such systems led them to seek out remote, isolated environments that were persistently stable, complex, and biodiverse. As they explain in "Trophic Structure and Productivity of a Windward Coral Reef Community on Eniwetok Atoll," the problem was that "mankind's great civilization is not in steady state and its relation with nature seems to fluctuate erratically and dangerously."[28] They viewed human impacts as a major disruption to the ways that systems self-organized and regulated. Framing environments as systems required removing humans from the already complex equation. With large amounts of funding supplied by the Atomic Energy Commission, the Odums, and many other ecological scientists, conducted research in the Marshall Islands and at other sites where nuclear testing provided "a unique opportunity [...] for critical assays of the effects of radiations due to fission products on whole populations and entire ecological systems in the field."[29] Thus, nuclear pollution presents a *kairos* of rhetorical opportunity and opportunism for the energy rhetoric of ecosystems (which I unpack in Chap. 5). Because of this, as Martin puts it in her groundbreaking article, "…ecosystems cannot be understood apart from the history of Cold War nuclear violence."[30] This history of ecological systems research, deeply rooted in Cold War geopolitics, helps foreground how ecosystems produce a version of "Nature," similar in some ways to what William Cronon examines in his critique of "wilderness" as a social construct.

These radioecology experiments represent some of the more controversial aspects of Odum's research, particularly focusing on his studies in the Marshall Islands and Puerto Rico, which were sites of significant nuclear testing and military activity in the mid-twentieth century. This discussion sets up the examination of ecological problems that persist in shaping ecological inquiry today, topics that are expanded on in Chaps. 4 and 5. Odum's extractive research at these sites is a stark example of the ways that extractive science perpetuates colonialism, as Max Liboiron argues in *Pollution Is Colonialism*. As Liboiron explains, scientific practices that rely on "entitlement to Land" that assume the right to create and distribute pollution are part of the "bad relations" that constitute colonialism.[31] Building from Danielle Endres' recent work with nuclear decolonization and Indigenous resistance, and by focusing on the specific context of nuclear colonialism and Odum's research, this chapter critiques ethical and political dimensions of extractive scientific practices and demonstrates their connections to ecology and ecosystems. At these research sites, Odum found a "proving ground" for the ecosystem theory, both in the sense of an environment where a theory can be tested, and in the sense of a testing place for military technology.[32] Building from the previous discussion of energy, this chapter focuses on how the energy of nuclear radiation was used to help invent the ecosystem concept.

The context of nuclear colonialism and extractive practices contributes to the twin spatiotemporal problems discussed in Chaps. 4 and 5, problems which leave ecology "out of place" and "out of time." The rhetoric of energy deployed by Odum's macroscopic view rendered ecosystems abstracted in both place and time. These spatial and temporal "slices"[33] or "cuts"[34] are used to reduce complexity across a closed system. This produces a seductive, yet ultimately misleading, image of ecology, which is fixated on the ever-emerging present moment and renders place as an abstraction that is removed from historical, political, and cultural context. For example, Raby explains that "the project's apparent secrecy generated suspicion within nearby communities," and years in the future "the area remained a hotspot for tales of secret military science experiments, encounters with aliens, and the legendary chupacabras, as well as protests over logging and unexplained park closures."[35] Examining historical and geopolitical elements of Odum's research reveals how colonialism persists in shaping both modern ecological science and rhetorical inquiry. Observing the long-term environmental and social impacts of militarized nuclear colonialism helps contextualize and situate Odum's resulting radioecology studies in the Marshall Islands and Puerto Rico. As such, this chapter serves as a bridge between the concept of energy discussed in Chap. 2 and the concepts of scale and temporality discussed in the following chapters. These problems of space and time, as this chapter sets up, are deeply rooted in the context of nuclear colonialism.

COLONIAL ECOLOGIES OF NUCLEAR POLLUTION

These and other examples demonstrate how, from WWII through the Cold War, ecology was enfolded in the project of militarized nuclear colonialism. In *On the Frontier of Science*, Leah Ceccarelli examines the rhetoric of the scientific frontier, critically examining how frontierism frames inquiry as both exploration and exploitation. She describes how scientific advancement is often linked with the type of manifest destiny associated with the type of frontier nostalgia that Cronon critiques. Her insights help illuminate how H. T. Odum's fieldwork in Puerto Rico and the previous work of both H. T. and Eugene Odum in the Marshall Islands, conducted in partnership with the US Atomic Energy Commission, in many ways exemplifies the exploitation of vulnerable ecosystems and populations in the name of scientific discovery. As Ceccarelli notes, frontier rhetoric "justifies the expansion of control over new territories under the guise of progress," a dynamic clearly at play in these nuclear proving grounds.[36] The history of radioecology reflects the extractive and opportunistic frameworks of frontier science, where humans and the environment are exploited in the name of progress and in staking claim to uncharted areas of knowledge. In Chap. 5, I will discuss how the concept of *kairos* has been applied to understand how scientific discovery operates on an opportunistic temporality based on "*growth* and *progress.*"[37] For our purposes here, it is enough to say that ecosystems relied on the kairotic metaphors of frontier science, opportunity, and progress to justify harmful colonial relations to lands and peoples. In turn, radioecology was supported by the AEC as part of broader efforts to understand, and justify, the use of nuclear technology.

Along similar lines, Max Liboiron argues in *Pollution Is Colonialism* that scientific research perpetuates colonial practices by treating land and ecosystems as resources to be extracted, controlled, and polluted without considering Indigenous sovereignty or long-term impacts on local environments and communities. Liboiron explains that science often maintains exploitative practices that are normalized because "[c]olonial land relations are inherited as common sense, even as good ideas."[38] The presumption of access to land, and the privileged right to pollute and destroy that land, is part of what Liboiron focuses on in their critique of extractive frameworks of colonialism in science. The Odum brothers' research in Puerto Rico and the Marshall Islands serves as a stark example of this form of scientific exploitation. By treating these islands as experimental "proving grounds" for the ecosystems concept, their research reinforces the colonial power dynamics that subject Indigenous peoples and Lands to nuclear violence in the name of scientific progress and is thus "an enactment of ongoing colonial relations to Land."[39] In similar fashion to the closed dynamics of Silver Springs (discussed in Chap. 2), the presence of radiation in these environments produces a controlled setting in which researchers could observe how radiation impacted systems and energy flows. Radioactive pollution transformed the natural world into a living laboratory, where the concept of ecosystems as self-regulating networks could be put to the test. Here, radiation, pollution, and cybernetics converge in the production of a method for legitimizing the ecosystem concept as product and process of colonial Land relations.

Building from Liboiron, Danielle Endres' brilliant book *Nuclear Decolonization: Indigenous Resistance to High-Level Nuclear Waste Siting* examines how Indigenous groups have successfully resisted nuclear colonialism, detailing how the Western Shoshone, Southern Paiute, and Skull Valley Goshute peoples and nations successfully halted the development of nuclear waste sites on their land and setting forth "an Indigenous theory of social change."[40] Endres critiques nuclear colonialism by engaging with Indigenous scholars who "describe how convergences of systems of nuclearism and settler colonialism disproportionately harm Indigenous peoples, their Lands, and their lifeways throughout the life cycle of nuclear technologies."[41] Endres offers an important, and often overlooked, perspective on Indigenous rhetoric and/as acts of ongoing survivance and resistance to colonialism in all of its protean forms that goes beyond a "damage-centered research approach."[42] Instead, Endres celebrates the ongoing lived experiences and acts of resistance to nuclear colonialism by Indigenous activists. Similar motivations animate the final section of this book, which seek to move beyond treating history as abstract or purely theoretical, and toward understanding how this history produces an exigency that motivates us toward different types of methods and practices that can help us support and contribute to these anti-colonial acts of resistance.

THE RHETORIC OF SCIENCE MEETS THE RHETORIC OF TECHNOLOGY

While the rhetorics of science and technology are often framed as distinct lines of inquiry,[43] these foci are deeply entwined in ecosystems ecology. This convergence of technosphere and biosphere reveals how scientific and technological frameworks mutually shape how we perceive and act upon the natural world. Together, the rhetoric of science and technology functions to normalize extractive colonial approaches to scientific research. For example, drawing metaphors from cybernetics and economics fundamentally altered how researchers in the sciences and humanities conceptualized the flow of energy and/or information within systems. The Odum brothers' use of cybernetics positioned ecosystems as self-regulating machines that could be understood and managed through technological intervention, viewing ecosystems as input-output systems. As computers, networked modeling, and feedback loops became part of the rhetoric of ecosystems science, the lines between science and technology became blurred. The incursion of the technosphere into the biosphere was characterized by extractive methodologies that treated the natural world as a resource to be interfaced, controlled, and optimized for human benefit. What began as a theoretical and linguistic endeavor quickly had material impacts on local environments and the peoples who called those places home.

In places like Puerto Rico and the Marshall Islands, where ecosystems research was conducted under the guise of scientific progress, technological tools and innovation were used by the US military to justify the displacement of Indigenous peoples and the exploitation of natural resources. As I will discuss in Chaps. 4 and 5, these sites of research were not just locations for scientific inquiry, but also spaces where the rhetoric of technology was used to obscure the colonial realities of scientific work and military interests. The rhetoric of technological advancement deflected critiques of the damage wrought by nuclear testing, both to environments and to the people who lived there. As such, the ecosystem itself became a rhetorical tool that justified displacement and exclusion of peoples from their traditional lands at sites such as Bikini and El Verde. Attending to the complex history of these lands requires ecological inquiry to account for human impacts across vast expanses of time and space. As such, the use of ecology as a framework for RWCS makes requisite such land-based perspectives, calling us to work toward more just and equitable projects that meet the highly situated needs of both human and nonhuman communities. In order to address the legacies of colonialism and extractive research, we must first understand the ways that the rhetoric of science and technology shapes ecological inquiry.

Islands in the Scheme

It is not merely a coincidence that much of the Odum brothers' research took place in seemingly remote places that were iconically linked to notions of "Wilderness" and "Nature." Rather, Nicole Starosielski's *The Undersea Network* offers an important basis for understanding how the conception of islands as insulated from global society connects to their role in making digital networks possible at the global scale.[44] As such, her book is an important premise to understanding how ecosystems participate in linking prior notions of nature and wilderness with the technics of digital networks. Starosielski discusses how islands, though considered geographically remote and isolated, are key nodes in the telecommunication infrastructure, comprising the historic pathways of telegraph cables to the fiber optic cable networks that today carry the bulk of transcontinental Internet traffic. Islands support the interconnectivity of the network, joining vast and disparate places together. As sites of cable landings, the islands facilitate the flow of information, but they are in turn connected through that infrastructure to the global technosphere. Similarly, islands were vital to ecosystems research, particularly in the work of Eugene and H. T. Odum. Just as they serve as vital hubs for telecommunication networks, so do they also function as crucial proving grounds for the move from ecosystems as an abstract theoretical concept to the materialization of methods and practices for ecosystems ecology.

Beginning in the 1950s with sites like Silver Springs, Florida (as discussed in the previous chapter), and expanding with studies at Enewetak Atoll in the Marshall Islands and the El Verde research site in Luquillo, Puerto Rico, the ideas of isolation and remoteness were crucial for producing the closed dimensions necessary to study the trophic dynamics of ecosystems. As I expand on in the next two chapters, the purportedly rigid boundaries of these environments allowed ecologists to reduce some of the complexity that often made large-scale ecological research impossible. These controlled environments provided an ideal place to study energy flow in ways that mirrored the cybernetic feedback loops of information and systems theories. Yet, as Starosielski demonstrates, the idea of islands as isolated is illusory, as islands are, and have always been, connected to global ecologies and technological systems. In addition to their role in the colonial infrastructure of telegraph cable networks, these islands were also important locations for the Pacific theater of WWII. The Odums' work reflected this duality: while ostensibly studying the local environment, their research connected those individual sites to abstract and global theories of ecological systems, much like the undersea cable networks join remote locations to a global communications network. This context also demonstrates the depth of spatial and technological connections between war, nuclear colonialism, information theory (born from WWII communications work), digital networks, and ecosystems ecology at sites like Enewetak, where these different elements converge and wrap around the island like the tentacles of a cephalopod.

The role of "networked islands" in global communications can also help us understand how cybernetic, technocratic, and colonial infrastructures shaped the invention and development of ecosystems ecology. Many of the key sites where fieldwork in ecosystems ecology took place were located near cable landings, demonstrating the entanglement between these islands not only as examples of "Wilderness" or "Nature," but also as convergence points between technosphere and biosphere. The proximity of these ecological research sites to telecommunication nodes reflects a deep network of connections between biosphere, semiosphere, and technosphere. In both cases, the islands are not merely isolated environments but are integral parts of larger, global infrastructure of knowledge and control systems. Thus, these sites served as "proving grounds" not only for technological connectivity and nuclear testing but also for ecological theories that would later be applied globally. The metaphor of "islands" as isolated ecosystems parallels the way these same islands function in global communication systems. The very notion of isolation becomes a rhetorical tool for understanding both technological and ecological systems as contained, manageable, and ultimately controllable. However, as Starosielski's work and the Odums' research show, these islands are deeply connected to global systems, whether through undersea cables, ecological energy flows, or nuclear pollution. This connectivity challenges the myth of isolation and reveals how islands are central to both global communication networks and ecological research, serving as key nodes in the circulation of information and energy that define both systems.

Notes

1. William Cronon, "The Trouble with Wilderness; Or, Getting Back to the Wrong Nature," in *Uncommon Ground: Rethinking the Human Place in Nature*, ed. William Cronon (W. W. Norton & Company, 1995), 80. Emphasis added.
2. For a compelling land-based perspective on the history of the El Yunque, see Megan Raby, "Slash-and-burn ecology": Field science as land use. *History of Science*, 57, no.4 (2019): 441–468, https://doi.org/10.1177/0073275318819656
3. María Uriarte, Jill Thompson, and Jess Zimmerman, "Hurricane María Tripled Stem Breaks and Doubled Tree Mortality Relative to Other Major Storms," *Nature Communications* 10 (2019), doi:10.1038/s41467-019-09319-2
4. Ariel Lugo, "H.T. Odum and the Luquillo Experimental Forest," *Ecological Modelling* 178 (2004): 67, doi:10.1016/j.ecolmodel.2003.12.023
5. For example, the oncologist Cornelius Rhoads wrote a letter in 1931 espousing racist views and stating that he infected Puerto Rican human subjects with cancer cells. His letter was shared with the American media and caused a scandal, but he never faced consequences in his lifetime. His name was later removed from an award given by the American Association for Cancer Research in 2003. See Douglas Starr, "Revisiting a 1930s

Scandal, AACR to Rename a Prize," *Science* 300, (2003): 573–574, doi:10.1126/science.300.5619.573. Along the same lines, in 1956, Gregory Pincus and John Rock conducted the first human trials for contraceptive pills in Río Piedras. See Suzanne White Junod and Lara Marks, "Women's trials: the approval of the first oral contraceptive pill in the United States and Great Britain," *J Hist Med Allied Sci.* 57, no 2 (2002):117–160, doi: 10.1093/jhmas/57.2.117. This unethical study subjected the women to dangerously high levels of Enovid without informing them of the risks involved. In addition to these controversies and scandals from the medical community, the US Navy used the island of Vieques as a proving ground for weapons testing, including napalm and Agent Orange. At the same time, the military was also conducting experiments with mustard gas on enlisted men in WWII, with studies led by Rhoads that specifically targeted Black and Puerto Rican troops. See S. L. Smith, "Mustard Gas and American Race-Based Human Experimentation in World War II," *Journal of Law, Medicine & Ethics* 36, no. 3 (2008): 517–521, doi:10.1111/j.1748-720X.2008.299.x
6. Cronon, "The Trouble," 10.
7. Ludwig Wittgenstein, *Philosophical Investigations* (Blackwell, 1953).
8. Cronon, "The Trouble," 12.
9. Ibid., 18.
10. Ibid.
11. See H. T. Odum and R. F. Pigeon, eds., *A Tropical Rain Forest: A Study of Irradiation and Ecology at El Verde, Puerto Rico* (U. S. Atomic Energy Commission, 1970).
12. See Laura J. Martin, "Proving Grounds: Ecological Fieldwork in the Pacific and the Materialization of Ecosystems," *Environmental History* 23, no. 3 (2018): 567–592, doi:10.1093/envhis/emy007
13. See Raby, "Slash-and-burn ecology."
14. Eugene Odum, "The Emergence of Ecology as a New Integrative Discipline," *Science* 195 (1977): 1290, doi:10.1126/science.195.4284.1289
15. Aimee Bahng, "The Pacific Proving Grounds and the Proliferation of Settler Environmentalism," *Journal of Transnational American Studies* 11, no. 2 (2020): 57, doi:10.5070/T8112049580
16. Donald Worster, *Nature's Economy: A History of Ecological Ideas* (Cambridge University Press, 1994), 364.
17. Cronon, "The Trouble," 19.
18. H. T. Odum and B. Odum, "Concepts and Methods of Ecological Engineering," *Ecological Engineering* 20, no. 5 (2003): 339, doi:10.1016/j.ecoleng.2003.08.008

19. Timothy Morton, *Ecology without Nature* (Harvard University Press, 2007), 22.
20. Raby, "Slash-and-Burn," 461.
21. Morton, *Ecology without*, 103.
22. Dana Phillips, *The Truth of Ecology: Nature, Culture, and Literature in America* (Oxford University Press, 2003), 63.
23. N. Katherine Hayles, "Making the Cut: The Interplay of Narrative and System, or What Systems Theory Can't See," *Cultural Critique* 30, no. 1 (1995): 71–100, doi:10.2307/1354433
24. N. Katherine Hayles, *How We Became Posthuman: Virtual Bodies in Cybernetics, Literature, and Cybernetics* (University of Chicago Press, 2008), 12.
25. Kathleen Dean Moore, *How It Is: The Native American Philosophy of V.F. Cordova* (University of Arizona Press, 2007), 51.
26. Dan Ehrenfeld, "'Sharing a World with Others': Rhetoric's Ecological Turn and the Transformation of the Networked Public Sphere," *Rhetoric Society Quarterly* 50, no. 5 (2020): 309, doi:10.1080/02773945.2020.1813321
27. Rachel Wolford, "When A Woman Owns the Farm: A Case for Diachronic and Synchronic Rhetorical Agency." *Enculturation: A Journal of Rhetoric, Writing, and Culture* 22, no. 1 (2016). Accessed on February 2, 2025. http://enculturation.net/when-a-woman-owns-the-farm
28. H. T. Odum and Eugene P. Odum, "Trophic Structure and Productivity of a Windward Coral Reef Community on Eniwetok Atoll," *Ecological Monographs* 25, no. 3 (1955): 291, doi:10.2307/1943285
29. Ibid., 291.
30. Laura J. Martin, "Proving Grounds: Ecological Fieldwork in the Pacific and the Materialization of Ecosystems," *Environmental History* 23, no. 3 (2018): 569, doi:10.1093/envhis/emy007
31. Max Liboiron, *Pollution Is Colonialism* (Duke University Press, 2021), 5.
32. Martin, "Proving Grounds."
33. See Karen Barad, *Meeting the Universe Halfway: Quantum Physics and the Entanglement of Matter and Meaning* (Duke University Press, 2007).
34. See Hayles, *How We*.
35. Raby, "Slash-and-burn ecology," 466.
36. Leah Ceccarelli, *On the Frontier of Science: An American Rhetoric of Exploration and Exploitation* (Michigan State University Press, 2013), 32.
37. Carolyn Miller, "*Kairos* in the Rhetoric of Science," in *A Rhetoric of Doing: Essays Honoring James L. Kinneavy*, Steven P. Witte et al., eds. (Southern Illinois University Press, 1992), 314. Original emphasis.
38. Liboiron, *Pollution Is Colonialism*, 12.
39. Ibid., 6.

40. Danielle Endres, *Nuclear Decolonization: Indigenous Resistance to High-Level Nuclear Waste Siting* (Ohio State University Press, 2023), 2.
41. Ibid., 3.
42. Ibid., 5.
43. Carolyn R. Miller, "Opportunity, Opportunism, and Progress: *Kairos* in the Rhetoric of Technology," *Argumentation* 8 (1994): 81–96, doi:10.1007/BF00710705
44. Nichole Starosielski, *The Undersea Network* (Duke University Press, 2015).

Open Access This chapter is licensed under the terms of the Creative Commons Attribution-NonCommercial-NoDerivatives 4.0 International License (http://creativecommons.org/licenses/by-nc-nd/4.0/), which permits any noncommercial use, sharing, distribution and reproduction in any medium or format, as long as you give appropriate credit to the original author(s) and the source, provide a link to the Creative Commons license and indicate if you modified the licensed material. You do not have permission under this license to share adapted material derived from this chapter or parts of it.

The images or other third party material in this chapter are included in the chapter's Creative Commons license, unless indicated otherwise in a credit line to the material. If material is not included in the chapter's Creative Commons license and your intended use is not permitted by statutory regulation or exceeds the permitted use, you will need to obtain permission directly from the copyright holder.

CHAPTER 4

Ecology Out of Place: *Topoi* and Spatial Problems

Abstract Turning to the problem of place for environmental inquiry, this chapter examines the role of scale in ecosystems. It investigates the ways that both individual species and research sites are framed as discrete and self-contained units by drawing from scholarship in rhetorical theory and the environmental humanities. Scale plays an important role in framing ecosystems, from understanding processes like evolution to the relationship between global climate change and local environmental problems. Taken together, this chapter investigates how spatial rhetoric influences environmental inquiry.

Keywords Rhetoric of place • Scale • Topoi • Ecological engineering • Environmental management

> [P]lace convenes our being together, bringing human and nonhuman communities into the shared predicaments of life, livelihood, and land. Place calls us to the challenge of living together [...] Place calls us to the struggles of coexistence. —Soren C. Larsen and Jay T. Johnson, *Being Together in Place*[1]

Bay Campus, University of Rhode Island, Narragansett, Rhode Island, United States

© The Author(s) 2026
M. P. Jones, *Inventing Ecosystems*, Palgrave Studies in Media and Environmental Communication,
https://doi.org/10.1007/978-3-031-98793-9_4

92   M. P. JONES

The cool morning air seems to herald the end of summer as I stand watching my two-year-old daughter play on the beach. With the concentration of a surgeon, she is practicing packing damp sand into a plastic bucket and smashing it down to form a castle. Just beyond us, a crew is finishing construction on an update to the pier where the university's new Regional Class Research Vessel, the Narragansett Dawn, *will eventually dock. I glance up the hill to where the Bay Campus towers over us, scanning the various buildings where we occasionally hold department meetings and retreats, student defenses, and special events. Beyond the hill sits the Rhode Island Nuclear Science Center, home to the state's sole nuclear reactor. Behind the MERL (Marine Ecosystems Research Laboratory) building, I notice the mesocosm tanks off in the distance, and my thoughts turn to a thesis defense that took place a few weeks ago. The graduate researcher had been conducting a study on the Atlantic awning clam (*Solemya velum*) using the outdoor mesocosm tanks. The experiment ran into problems when an overabundance of a predatory annelid developed and likely killed off the clams, necessitating a pivotal change in the study. Standing on the beach, it suddenly occurs to me that this example lays bare the way that mesocosms reflect ecosystems more broadly, where complexity always seems to elude our grasp. Eugene Odum coined the term in a 1984 paper where he hoped to resolve divisions between "reductionism [and] holism," and to "bridge the gap" between "microcosms [and] macrocosms" as well as "between the laboratory and the real world."[2] Enter the mesocosm: a simulated outdoor environment that combines elements of the laboratory and the field, providing controlled conditions that allow researchers to study different types of effects in a replicable, regulated, and repeatable manner. In the paper, Odum directly references MERL as an example of the "middle size world" of the mesocosm.[3] At the same time, problems like the one encountered by the student demonstrate how incredibly difficult it is to simulate the dynamics of a real environment. While systems neatly package complexity into illusions of bounded models and metaphors, chaos resists containment. Just now, I feel a tiny hand touch my leg and look down to see my daughter grinning up at me from behind a seashell gripped between her thumb and forefinger.*

Locating Ecosystems

When we talk about an ecosystem, what exactly do we mean? What features distinguish the boundary of one system from another? What gets included and what is excluded? And what to do with the interstice, the ecotone? How can we separate environments into closed dimensions when the first so-called law of ecology tells us that "everything is connected to everything else"?[4] The answer has a lot to do with scale. From the Latin *scāla*, meaning "staircase," "sequence," or "ladder," *scale* describes a graduated series of values or levels that together compose a system. Scale is a tool that allows us to segment reality, calibrate perceptions, and quantify relationality. Without scale, it would be difficult to navigate or to use mental models to make predictions about the future. Using scale, we can demarcate boundaries and distinguish between different things, from objects to units to concepts. As such, scale has a rhetorical function in our everyday lives. From clocks to calendars to music to maps, scale helps us to demarcate and locate ourselves in time and space. Scale provides structure for us to categorize the world around us, but it also creates rhetorical complexities, particularly when we consider the sometimes arbitrary dimensions that characterize the spatial boundaries of ecosystems.

For the past few decades, numerous scholars have returned to the pride of place that spatial theory has long held in the field of writing, rhetoric, and communication studies (RWCS). For example, in rhetorical theory, the ancient concept of *topoi* (or *loci* in Latin), meaning roughly "topics" or "location," has long been a key point of connection between writing and geography (or γεωγραφία, a combination of "earth" and "writing"). *Topoi* are methods, premises, and assumptions that situate arguments within a particular field of inquiry.[56] As Christopher Keller and Christian Weisser put it in their introduction to *The Locations of Composition*, as a guiding concept, *topoi* call us to investigate how "places are located: how they relate to other places inside and outside the discipline, how our activities carve out new spaces from these places, and how these places allow us to alter, change, position, reposition, and move through our scholarly work and practices."[7] The rhetorical influence of *topos* can be mapped across RWCS, from its basis in ancient Greek rhetoric to its prominence in recent RWCS scholarship focused on spatial rhetoric, such as Nathaniel Rivers' "geocomposition,"[8] John Tinnell's *Actionable Media*,[9] Jacob Greene's *Composing Place*,[10] and most notably, Nedra Reynolds' *Geographies of Writing*.[11] Just as scale is a central element of topography— from the ancient Greek word τοπογραφία, which could be translated to place (*topos*) writing (*graphia*)—it is also a locus for contemporary RWCS scholarship that engages with place. For example, Reynolds builds from human geography research to understand scale as a key component of "how the social production of space stretches out from individual bodies to the home, neighborhood, city, region, nation, and world."[12] As such, *topos* offers a key concept connecting the spatial dimensions of writing to the scales at which persuasion and communication occur, from common ground to distanciation. At the same time, topologies can also serve to elide the rhetoric of scale, causing us to mistake the map for the thing it represents.

Scale and its Discontents

When we apply scale to understand information, we encounter what scientists call "scale effects," referring to the ways that models compare to the actual things they are meant to represent. In the fields of Science and Technology Studies (STS) and EH/DH, a growing area of research investigates "scale criticism," which applies insights drawn from the humanities and social sciences to understand the way scale shapes and limits inquiry. In discussing how ecosystems ecologists like Odum grappled with the complexities of studying ecosystems at different trophic levels, this chapter explores the ways that scale shapes how we perceive and navigate the world, from large-scale scientific studies of massive biomes to micro-scale studies that allow us to see the otherwise invisible. This chapter discusses the rhetorical challenges of scale, particularly in how systems are bounded and enframed in both study design and communication. The use of scale can reinforce hierarchical models that assume fixed boundaries and naturalize arbitrary distinctions. Scale can produce intractable problems when we mistake scaled boundaries as "natural" phenomena, such as in the case of the *scala naturae*, or the Great Chain of Being, which depicts a topological hierarchy that extends from the divine to the earthly—from god to angels to people, to animals, then to plants and nonliving things, and lastly to the denizens of hell (Fig. 4.1). Using a hierarchical scale, this model stratifies organic life and inorganic matter alongside the natural and preternatural to produce a general theory of spatial relations that crosses physical and metaphysical planes.

The roots of this philosophical tradition, at least in part, can be traced back to the Demiurge of Plato's *Timaeus*,[13] as well as his Theory of Forms and philosopher-kings in the *Republic*[14] and across other dialogues. Plato directly influenced Aristotle's characterization of the *scala naturae* in the *Historia Animalium*, which moves in the "upward scale" ascending from "lifeless things" through plants on a "continuous scale of ascent towards the animal," each measured by "graduated differentiation in

Fig. 4.1 A drawing of the *scala naturae*, or the Great Chain of Being, by Fray Diego de Valadés, *Rhetorica Christiana* (1597). Wikimedia Commons. Retrieved from https://commons.wikimedia.org/w/index.php?curid=33603873

amount of vitality and in capacity for motion."[15] Such a model produces an anthropocentric, either human- or god-centered scale, with a philosophical vitalism organizing the hierarchical gradations from perfection and complexity to imperfection and simplicity. For instance, consider Disney's 1994 film *The Lion King*, where the lionized carnivores are charged with maintaining balance across the savannah, from the grass to the antelope to the king himself. While such a hierarchy is no longer a dominant framework for understanding biotic and abiotic relationships in science, it is still a powerful force that shapes both cultural and scientific imaginations.[16] Today, scalar thinking shows up in evolutionary discourse in the form of progressive advancement, such as the idea of more "primitive" and "advanced" or "higher" and "lower" life forms, and in environmental sciences through the tensions between the "food chain" and the "food web."[17] For instance, popular understandings of humans as "the top of the food chain" reflect this (false) scalar imaginary.[18] Despite the ecological perspectives advanced in parts of *The Lion King*, its depiction of disruption and succession, both political and ecological, naturalizes the divine patriarchal monarchy as a stabilizing force in the Pride Lands of Tanzania.

Charles Darwin proposed evolution as an opposing theory to the Great Chain of Being, writing in his notebooks that it is "absurd to talk of one animal being higher than another."[19] At the same time, he forwarded an arboreal metaphor in *Origin of Species* to describe his phylogenetic visualization representing the evolutionary process (Fig. 4.2A).[20] His tree-like representations and metaphors were taken up by Ernst Haeckel who helped popularize them in his work with phylogeny (Fig. 4.2B), even as he misconstrued the positioning of man as the pinnacle of evolution (Fig. 4.2C).

Varying interpretations and misrepresentations of Darwin's theories in science, religion, and popular culture reproduced scalar perspectives on progression from simplicity to complexity. The *scala naturae* also informed white supremacist rhetoric in theories of Social Darwinism, placing humans on different racial scales.[21] Such rhetoric shows up in any inquiry that views species or groups as "higher" and "lower," but it also is a clear influence on the popular ecological concept of trophic mapping, which visualizes how energy moves up the food chain. H. T. Odum's famous study at Silver Springs relied on similar scalar dynamics to understand how energy moves through a closed system (Fig. 4.3).

4 ECOLOGY OUT OF PLACE: *TOPOI* AND SPATIAL PROBLEMS   97

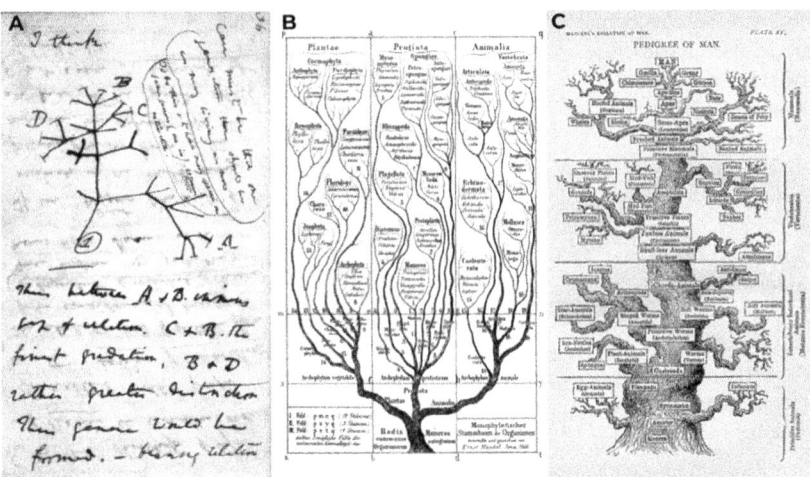

Fig. 4.2 An 1837 sketch by Charles Darwin (Fig. 4.2A) from his *First Notebook on Transmutation of Species*, visualizing an evolutionary tree, likely his first diagram representing evolution, courtesy of Wikimedia Commons. Retrieved from https://commons.wikimedia.org/wiki/File:Darwin_tree.png.A similar tree-like structure (Fig. 4.2B) was used by Ernst Haeckel in his 1866 book *General Morphology of Organisms* to visualize what he dubbed "phylogeny," or the evolutionary history of an organism, courtesy of Wikimedia Commons. Retrieved from https://commons.wikimedia.org/wiki/File:Haeckel_arbol_bn.png. His "Pedigree of Man" visual (Fig. 4.2C), published in his 1879 book *The Evolution of Man*, most clearly reproduces the *scala naturae*, courtesy of Wikimedia Commons. Retrieved from https://en.wikipedia.org/wiki/File:Tree_of_life_by_Haeckel.jpg

Trophic conceptual models can be useful in understanding energy flow, but they can also drastically shape how we perceive relationships within an environment. Odum's work with trophic mapping built on the prior work of Hutchinson[22] and Lindeman[23] who developed the idea of trophic levels based on Charles Elton's prior concept of a "pyramid of numbers."[24] Today, the trophic pyramid is a common representation of ecology in popular culture, reproducing scalar dynamics of "higher" and "lower" organisms and relationships. While Odum's engagement with Lotka's "maximum power principle" wasn't an overt element of his early work at Silver Springs, he directly discusses the principle of "trophic level efficiency," citing Lindeman and Dineen,[25] and lamenting that the model developed from the Silver Springs study was probably not accurate enough to put this

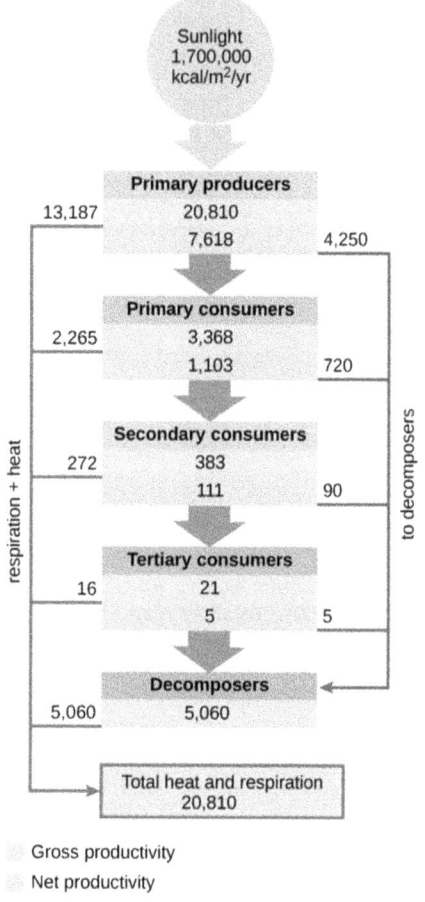

Fig. 4.3 A conceptual model used to describe the structure and dynamics of an ecosystem, based on H. T. Odum's Silver Springs study, courtesy of *Biology 2e* from OpenStax, licensed under Creative Commons Attribution License v4.0. Importantly, energy decreases with each order of magnitude

theory to the test.[26] The idea of the maximum power or hierarchical efficiency of complex systems reflects core tenets of the Great Chain of Being, with ecosystems presumably becoming more complex and dynamic over time until they reach a point of maximum efficiency à la *ens perfectissimum*, as well as characterized by a flow of energy from simple input to complexity as it ascends the trophic map.

Odum's 1950 dissertation, "The Biogeochemistry of Strontium: With Discussion on the Ecological Integration of Elements," was one of the earliest ecological studies that engaged directly with systems theory on a global scale,[27] anticipating perspectives of the Earth as a single organism or system decades before these ideas were popularized in James E. Lovelock and Lynn Margulis' "Gaia hypothesis."[28] Importantly, Odum's research contributed data-driven analysis to better understand how we might scale ecosystems to the planetary level. According to Karin Limburg, the topic was likely suggested by his dissertation director, G. Evelyn Hutchinson, renowned ecologist and limnologist,[29] who had previously published on a related subject.[30] Limburg notes that Odum's research "emphasized self-regulation, biotic–abiotic linkages, cybernetics, and steady state."[31] Odum further argues that "the strontium cycle seems to qualify as one of the large entities which in ecological literature are known as ecosystems" and suggests the "strontium ecosystem" to be a "proper application of the term."[32] This work prepared Odum for his later work mapping ecosystems at both large and small scales. With his early focus on ecosystems at such a scale, it is little surprise that it would figure prominently into his theories of ecosystems.

While scale is integral to ecological thinking, it can also pose limits on the way we conceptualize environments. Just as the ecosystem is conceptually rooted in scale, so the ecological crisis is, in part, rooted in a rhetorical problem of scalar displacement. Today, the environmental problems we face are of such magnitude that they overwhelm our ability as individuals to rhetorically situate or locate them, resulting in what Andrew Pilsch (building from Benjamin Bratton) refers to as an "invigorated nihilism" that is "emboldened by the impending doom of climate change, amok neoliberal economics, and increasingly brutal planetary-scale software ecologies."[33] In other words, systemic problems become intractable when confronted at the wrong scale(s). In order to elaborate on how scale defines relationships between and among rhetoric and ecological inquiry, this chapter turns to the concepts of the "biosphere" and the "ecotope," the latter of which was coined by Thorvald Sørensen in 1936 and further developed by Arthur Tansley in 1939 as part of the "basic units of nature" that helped produce the ecosystem concept.[34] An ecotope is the smallest distinct ecosystem and could be contrasted with the biosphere, which describes a global ecosystem. Similarly, H. T. Odum proposed the "macroscope" as a key tool for ecologists to eliminate details in ecosystem modeling,[35] while Eugene Odum coined the term "mesocosm" to describe an outdoor experimental setup that brings together the constrained elements of the laboratory with the complexity of the natural world.[36] In the following sections, I will define scale critique in EH, apply this concept to

Odum's work, and explore the lasting impact of this research by turning to the examples of Biosphere 2 and the US space program.

Scale Critique

In their pathbreaking book, *Being Together in Place: Indigenous Coexistence in a More Than Human World*, Jay T. Johnson and Soren C. Larsen draw from both Western phenomenology and Indigenous knowledges to understand place as an agential force. They demonstrate how, "...[b]eginning in earnest with the Enlightenment, European political discourse began to construct autonomy as an abstract, exclusive right vested in a singular, secular political subject—citizen and public—creating a new scale of state authority."[37] Through Indigenous knowledge practices, they resist the "hegemonic, hierarchical, and oppressive scale" of "state sovereignty"[38] to define "an active agency of place."[39] This agency binds humans and nonhumans in "a way of being and knowing" which they refer to as "scales of coexistence."[40] Building from Vine Deloria, Jr. and Daniel Wildcat, they resist the derangement of place through scale, arguing that "[t]o be Indigenous means 'to be of a place.'"[41] Following their claims about scale and sovereignty, this section seeks to critique scale as a colonial method for mapping relations.

In an overlapping fashion, the environmental crisis is, on the one hand, a global phenomenon, and on the other, made up of mundane, individual actions. In the face of such massive environmental problems, our very sense of individual agency seems to vanish. This incongruity is an example of *the problem of scale*, an issue pored over by posthumanists, ecocritics, and STS scholars[42] that has recently garnered interest from rhetoricians.[43][44] Rhetorically, scale invokes Aristotle's concept of magnitude (*megethos*), which, as Jenny Rice demonstrates, describes the ways that "abundant information accumulates in ways that expand beyond epistemic registers, creating a sense of coherence."[45] Through these orders of magnitude, scale shapes our experience with place and relationality. While mapping technologies like Google Maps reinforce the impression that we can neatly zoom from one scale to another, such changes in scale produce ontological rifts that necessitate changes in subjectivity, agency, and ethics that don't necessarily fit within human(istic) models. Scales separate places into nested dimensions that seem to zoom neatly between micro and macro, but as numerous scholars demonstrate, different levels of scale require ways of thinking that subvert basic tenets of humanism (such as agency).

Yet, as ecocritic Zach Horton demonstrates, scale is a fundamental element of ecology. Horton argues that the "scales we isolate [are] a matter of narrative framing."[46] For ecology, scale is "an enabling fiction" that allows scientists to apprehend, map, and visualize relationships across complex systems.[47] Just as scaling becomes a practical tool for ecological modeling, so does it introduce a philosophical dilemma, demanding that we confront the limits that our narrative frames place on our perception and understanding of ecological dynamics. Scale critique enables us to understand phenomena from different scalar perspectives.

As part of scale critique, it is important to highlight the role *scale variance* plays in the rhetorical dimensions of ecosystems. Derek Woods introduces the concept of scale variance as part of critiques of the conventional view that ecosystems can be smoothly mapped across different scales. Rather, each level of scale—whether local or global, micro, meso, or macro—has distinct constraints, and understanding these discontinuities is essential to accurately representing ecological relationships. In programming and statistics, scale is not typically assumed to zoom in a seamless or continuous fashion and will always produce discontinuities. Akin to Bruno Latour's "anti-zoom" and Horton's critiques of "freescaling," scale variance challenges the basic assumption that agency remains stable across different scales and encourages scale literacy. It increases our understanding of the ways that information from one scale cannot be easily made to interface with data from another. The phenomenon of scale variance makes clear that scale is not a neutral measurement tool separate from the bias of other storytelling techniques. Woods demonstrates that, in both ecology and in EH, scale is often presented as a seamless continuum, oversimplifying the complex ways that scale shapes the ways we define ecosystems. For example, climate change occurs at the scale of the planet through ubiquitous carbon emissions, each occurring locally but with global effects in aggregate. Aggregation does not necessarily operate on a 1:1 scale (i.e., with a straightforward causal relationship), but rather, it produces what Woods describes as "disjunctures," or breaks between scales. Different places are being impacted by climate change differently. The technomorphism of scale can serve to obscure the complex scales of ecosystems, often reinforcing anthropocentric biases and the ways that scale functions as a storytelling tool.

Scale and Ecosystems Ecology

In ecological science, scale is a fundamental structure that defines what kinds of relations constitute objects of study. Scale determines where a study takes place, producing the necessary dimensional boundaries to study ecosystems. Mesocosms mediate micro and macro scales between parts and wholes. Scale equips ecologists with dimensions that separate micro, meso, and macro levels, from a handful of dirt to an entire planet. Essentially, just as synchronic/*kairotic* time allows ecosystems ecologists to study a cross-section of time (as I discuss in the next section), so do topological scales permit scientists to study a cross-section of space. Scale helps produce systems rhetorically through spatiotemporal divisions, but in doing so, it also organizes dimensional levels into a hierarchical topology. As Dana Phillips puts it, "Ecologists were beginning to do macrobiology and fieldwork at the very moment when other scientists had become convinced both of the primacy and, more important, of the practicality and greater utility of microbiology and laboratory experiment."[48] For example, scientists might compare data taken at one scale, such as of an individual fish species, to data at a larger scale, such as a population of a certain riverbank or even riparian zones more generally. When referring to the "scale of species," scale abstracts the spatiotemporal dimensions of place to produce a topological hierarchy.

These abstractions enable ecologists to study vastly complex models, but these cuts also produce spatial and temporal paradoxes for ecology. One such problem is known as "Simpson's paradox," which refers to a statistical phenomenon in which a trend appears one way across multiple data sets but changes or disappears completely when those sets are combined. Recently, ecologists have suggested that this paradox (as it specifically occurs in ecology) is a direct result of assumptions made by limnologists involving spatiotemporal scale.[49] By relying on a large data set of sample averages, these studies often lose sight of some important factors that define and influence individual lake ecosystems, what ecologists refer to as "confounders." To address issues like these, contemporary ecologists employ a cross-scalar approach to ecosystem modeling that builds models from large data sets but also attempts to resolve confounders by taking site-specific factors into account. Yet, even with these emerging methods, scalar abstraction remains an essential component of ecosystems ecology, dating back at least to Howard T. Odum's influential "Silver Springs

Model" (Fig. 4.4), developed from the first-ever comprehensive study of energy moving through a closed system.[50]

Through trophic mapping, Odum created what he would later refer to as a "macroscopic" approach to studying ecosystems holistically.[51] Macroscopic models are tools that "cut through the plethora of detail" and reduce complexity and render a holistic perspective, which Odum compares to the impressions of an artist. H. T. Odum's macroscope functions as a "detail eliminator" (Fig. 4.5), a tool for producing the spatial cut that makes it possible for ecologists to scale above the complex individual parts of a system.[52]

The macroscope allowed ecology to advance from a discipline that relied primarily on simplistic models of ecosystems to one that could

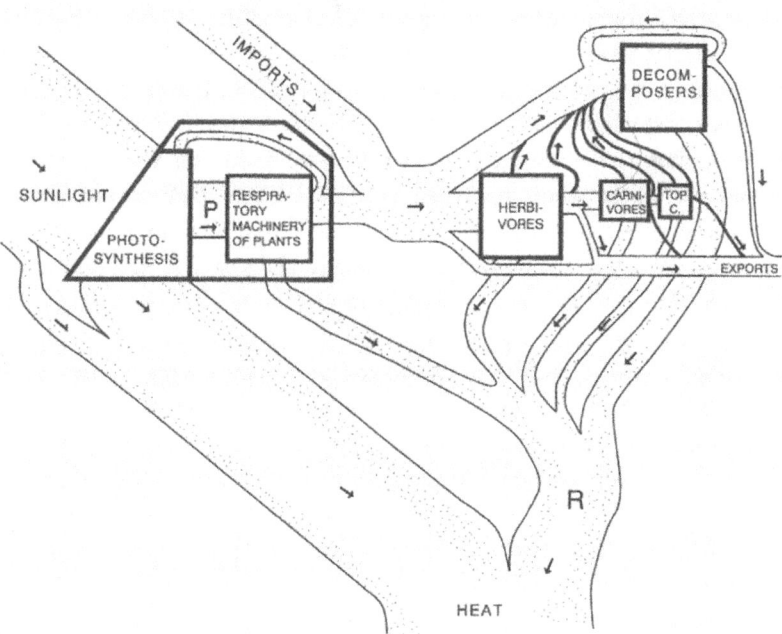

Fig. 4.4 "Silver Springs Model" from Odum's *Environment, Power and Society* (1971), courtesy of Wikimedia, https://commons.wikimedia.org/wiki/File:Silver_Spring_Model.jpg. The model represents herbivores, carnivores, and decomposers, as well as the flow of energy through the system

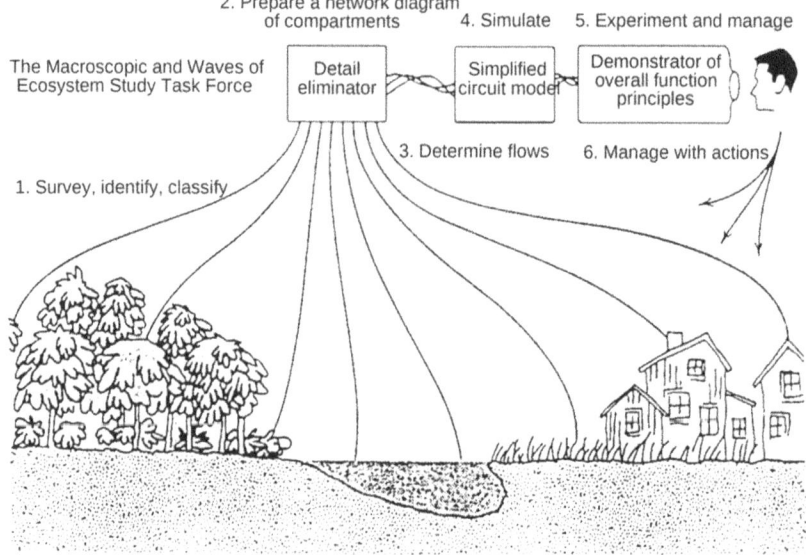

Fig. 4.5 Illustration of the macroscopic perspective from Odum's *Environment, Power and Society* (p. 10)

model complex systems based on in situ (or place-based) data. This method is directly implicated within ecosystems ecology's ties to nuclear colonialism, but just as this approach builds from this inheritance, it also obscures these connections. As previously discussed, macroscopic methodologies impose an abstract systems model onto the places they study, obscuring the autochthonous confounding factors that define specific places.

One of the early examples of this process was in the research that the Odum brothers conducted at the Enewetak Atoll. As Laura J. Martin demonstrates, their studies reversed the traditional conception of fieldwork, where data collected is later used to build an abstract model. Rather, Martin argues that the "Odums did not 'discover' evidence of ecosystems at Enewetak; rather, they theorized Enewetak's coral reefs as ecosystems years before they arrived" and, in conducting their research, the pair "struggled to match the species and situations they encountered to their preconceived frameworks."[53] This method helped to transform Enewetak's environment into a mesocosm, a system that the Odums studied to discover their ecosystem model. In other words, their fieldwork was, in part, conducted to match place to model, rather than to produce the model based on the local data. While the mesocosm concept has since received criticism in ecology for its reliance on abstract simplicity,[54] it was this very quality that made mesocosms appealing for early field studies of ecosystems. This research would support the concept of the pyramid of productivity that had been forwarded by Lindeman and Hutchinson[55] and which was highly influential in both Odums' thinking.[56] Trophic dynamics provide a spatial slice that encourages one to imagine how ecosystems might seamlessly scale from the micro to the macro level, from local fieldwork up to the scale of the planet. In doing so, their fieldwork rendered the Enewetak ecosystem into a pristine place akin to the wilderness concept critiqued by Cronon. And yet, as Martin demonstrates, "[T]he atolls were neither isolated nor pristine."[57] As such, their work at the Pacific Proving Grounds demonstrates connections between *kairos*, the subject of the next chapter, and ecological approaches to rhetoric. While many rhetoricians discuss *kairos* as a place-based concept,[58] the prevailing tendency is to treat place as an abstract concept. In this way, rhetorical ecologies are traditionally limited by a reliance on frameworks inherited from ecosystems ecology. To theorize rhetoric using an ecological framework is to place it within specific locations. In this case, that place is the Marshall Islands and the nuclear colonial violence that the people and environment continue to survive and resist today.

The Machine Flops?

The University of Arizona's Biosphere 2 is a research facility located just outside Oracle, Arizona. It was constructed over four years, starting in 1987, and cost $180 million, funded by Ed Bass, a philanthropist who was born into a family oil fortune. In the 1970s, Bass met ecosystems ecologist John P. Allen at the Synergia Ranch, a countercultural community which Bass designed based on Buckminster Fuller's "Spaceship Earth" concept. Fuller forwards the concept in the 1969 book *Operating Manual For Spaceship Earth*, where he describes the Earth as a spaceship flying through space and offers recommendations for how to properly manage, steer, and sustain the ship. His work was a force that brought together systems theory with ecology, advocating for a radical restructuring of society around technology, economics, and education. His philosophy repackaged technocratic optimism in a manner that found purchase in both the radical hippy movement and in garnering support from the Department of Defense. Alongside the Odum brothers' research on mesocosms and ecosystems, Fuller's radical theories were a major influence on Bass' vision for Biosphere 2. This massive facility—featuring a manufactured tropical rainforest, an ocean complete with waves and coral reef, a savanna, a mangrove wetland, a desert, a micro-city, and an organic farm—resembles something out of David Lynch's 1984 film *Dune* with its impressive structure of glass and steel sprawling out over three acres of the Arizona desert. The hermetically sealed environment of this human-scale terrarium was intended to support human life, allowing for closed study of ecological systems. Modeled on the systems of Biosphere 1 (Earth), Biosphere 2 sought to reproduce earth's life support systems in a micro scale, reducing its complexity to more simplified inputs and outputs, alongside a well-ordered trophic hierarchy that ecological operators could control.

Biosphere 2: Mission 1 was a massive experiment that ran from September 26, 1991 until September 26, 1993, in which eight researchers maintained permanent residence in the closed off structure while conducting research on human and environmental interaction. Like E. M. Forster's "Machine," as discussed in the first chapter, the Mission 1 experiment was plagued by a myriad of problems, from a wide variety of miscalculations to disagreements among the scientists inhabiting the research facility. The complexity of environmental systems threatened to overwhelm the experiment, with miscalculations creating a need to artificially scrub CO_2 and inject oxygen into the enclosure at various points in the process. Trees suffered from a lack of stress wood, where the natural pressure of wind helps to strengthen the tree as it develops. The artificial ocean had to be meticulously maintained for pH levels and to scrub excess nutrients. While these examples demonstrate the role that complexity plays in natural environments, as well as the limits of ecosystem management, Odum maintained that the aims and goals of the experiment were misrepresented. Reflecting on Biosphere 2 as an example of a mesocosm, Odum notes that scale led to "tragic confusion" because while the "management process during 1992–1993 [...] was in the best scientific tradition [...] some journalists crucified the management in the public press, treating the project as if it was an Olympic contest to see how much could be done without opening the doors."[59] Despite Odum's views, the Fuller-inspired dome became a symbol of the failures of technocratic ecological thought.

WHITEY ON THE MOON

The geodesic design of the Biosphere 2 facility was based on Fuller's radical architecture as much as his revolutionary theories. While he didn't invent the structure, Fuller popularized geodesic domes as an essential part of his approach to countercultural revolution. The future, according to Fuller, was going to be spherical. The geodesic dome, alongside Fuller's complex lexicon, has been a mainstay in representations of technocratic solutionism in popular culture, from films like the 1976 *Logan's Run* to the 2004 television series *Lost*, in otherworldly eco/tech-horror like Stephen King's 2009 book *Under the Dome* and Jeff Vandermeer's 2014 book *Annihilation*, and it has been more forcefully lampooned in comedies from the 1996 film *Bio-Dome* to the 2007 *The Simpsons Movie*. While today these technocratic approaches to environmental management might remind us of the stuff of science fiction, the concept of a closed system persists in shaping how we understand environmental problems and their solutions. A particularly compelling critique of technosolutionism and technocracy can be found in Gil Scott-Heron's spoken word poem, "Whitey on the Moon," which was released on his debut 1970 album, *Small Talk at 125th and Lenox*. In the poem, Scott-Heron juxtaposes the space race with the difficult lives of people on the ground, demonstrating inequities of funding for the space race. Just as Scott-Heron's critique highlights how technocratic solutions ignore and obscure social inequalities, contemporary examples of space exploration and technocracy abound. These examples point to the interstellar scales of injustice perpetuated by technosolutionists.

Along these lines, Casey Boyle's "A Wealth of Realities" discusses the enabling fictions deployed by Elon Musk in "Starman"—where a cherry-red Tesla was launched into space as a test payload for SpaceX's Falcon Heavy rocket—an "experiment that doubled as a publicity stunt."[60] Boyle examines the discourse around what many now call "the era of NewSpace," which is, in many ways, a direct inheritor of Bucky Fuller's imaginary. Boyle turns to Sun Ra and Afrofuturism for examples of possible alternatives to contemporary technosolutionism. Among a coterie of contemporary aspiring technocrats, Musk embodies Fuller's technocratic optimism *par excellence*. The Starman stunt generated "a daily dose of spectacle, an aesthetic of world-building rhetoric" that served to "remind us that human's self-appointed stewardship of this world has failed."[61] Musk's desire to colonize Mars likewise immediately must reckon with issues like

Spaceflight Associated Neuro-ocular Syndrome, where longer time spent in microgravity has extremely damaging effects on the body. Along the same lines, Jeff Bezos founded Blue Origin to build places to live and work in low Earth orbit, creating commercial destinations such as "Orbital Reef," a project that is "architected as a mixed-use business park 250 miles above Earth."[62] These plans read like something straight from J.G. Ballard's short story "Deep End," where humans seek to escape an earth they poisoned.[63] As these contemporary technocrats work to realize Fuller's "Spaceship Earth" (with them at the helm), they must confront the complexity that would come with reproducing the conditions of life on earth. Blue Origin flights have recently brought celebrities like William Shatner and Katy Perry into space—and into the news—met with vitriol and ridicule by the public. Similarly, Biosphere 2 stands as a monumental example of the technocratic optimism that underpinned the enabling fictions, grand aspirations, and inherent contradictions that were baked into attempts to engineer ecosystems as controlled, closed-off systems. As such, the failures of Biosphere 2 offer an instructive case study in the failures of technocratic optimism to understand the complexity of scale in ecology.

The *ethos* of technocratic solutionism that underpinned Biosphere 2 was as much a spectacle as it was a scientific endeavor. Like Fuller's vision of "Spaceship Earth," the facility presented an optimistic fiction of human mastery over ecological systems—a self-contained mesocosm that could theoretically be replicated on other planets. The technocratic spectacles that Boyle identifies in his critique of Musk serve to distract us from the widespread failures of environmental management here on earth. While Biosphere 2 was heralded as a groundbreaking experiment—one that was thought to hold the possibility for colonizing planets that currently don't support human life—the project struggled with a myriad of issues that similarly plague all such technosolutionist projects. On the whole, oversimplification of complex system dynamics brought unforeseen consequences. For example, while the trees grew faster inside the sealed dome, many collapsed under their own weight. While Biosphere 2 was an ambitious and pathbreaking project, it required numerous types of human interventions during the two-year experiment and was plagued by declining oxygen levels, overgrowth of particular species, and interpersonal conflicts among its human inhabitants. Like the mesocosm example at the beginning of this chapter, the problems encountered in Biosphere 2 demonstrate larger limitations to applying ecosystems models to both human and nonhuman problems. The aesthetics of Biosphere 2, inspired by Fuller's geodesic domes, continue to inspire a sense of futuristic environmental possibility. Yet, it also serves as a spectacle that obscures the messy realities of complex systems. As such, Biosphere 2 serves as a cautionary tale for applying scaled models and technocratic frameworks to environmental problems. Ecological inquiry must contend with the wide variety of scales through which we invent ecosystems.

Notes

1. Soren C. Larsen and Jay T. Johnson, *Being Together in Place: Indigenous Coexistence in a More than Human World* (University of Minnesota Press, 2017), 1.
2. Eugene Odum, "The Mesocosm," *BioScience* 34, no. 9 (1984): 558, doi:10.2307/1309598
3. Ibid.
4. See Dana Phillips, *The Truth of Ecology: Nature, Culture, and Literature in America* (Oxford University Press, 2003), 581. Phillips discusses this quote as an example of the type of poetic concepts that ecologists have worked to divest from over the past fifty years.
5. For instance, Aristotle discusses *topoi* in the *Topics* and the *Rhetoric*, distinguishing between general and commonplace *topoi*, where some topics are suited to specific disciplines (e.g., physics vs. ethics) while others distinguish between areas within those disciplines (e.g., deliberative vs. epideictic forms of rhetoric). Aristotle, *The Complete Works of Aristotle, Volume One: The Revised Oxford Translation*, ed. Jonathan Barnes (Princeton University Press, 1985).
6. For an in-depth study of the relationship between *topoi* and public opinion in social media, see Caddie Alford, *Entitled Opinions: Doxa after Digitality* (University of Alabama Press, 2024).
7. Christopher J. Keller and Christian R. Weisser, eds., *The Locations of Composition* (State University of New York Press, 2007), 5.
8. Nathaniel Rivers, "Geocomposition in Public Rhetoric and Writing Pedagogy," *College Composition and Communication* 67, no. 4 (2016): 576–606, https://doi.org/10.58680/ccc201629614
9. John Tinnell, *Actionable Media: Digital Communication Beyond the Desktop* (Oxford University Press, 2017).
10. Jacob Greene, *Composing Place: Digital Rhetorics for a Mobile World* (University Press of Colorado, 2023).
11. Nedra Reynolds, *Geographies of Writing: Inhabiting Places and Encountering Difference* (Southern Illinois University Press, 2004).

12. Ibid., 54.
13. Plato, *Plato: Timaeus and Critias*, trans. A. E. Taylor (Routledge, 2013).
14. Plato, *Republic*, trans. Robin Waterfield (Oxford University Press, 2008).
15. Aristotle, *Historia Animalium*, trans. D'Arcy Wentworth Thompson (The Clarendon Press, 1910), 588b.
16. See Carolyn Miller and Molly Hartzog, "'Tree Thinking': The Rhetoric of Tree Diagrams in Biological Thought," *Poroi* 15, no. 2 (2020): 2, doi:10.13008/2151-2957.1290.They discuss the enduring rhetorical power of the *scala naturae*, examining how this concept became a dominant metaphor in Western thought and permeated early biological classification systems, persisting into the Enlightenment and beyond. They explain how the *scala naturae* rhetorically aligns life forms along a single axis of perfection, with humans near the top, creating a framework that conceptualizes the natural world as inherently ordered and stratified.
17. Miller and Hartzog discuss several examples from contemporary biology. For example, E. Rigato and A. Minelli, "The great chain of being is still here," *Evolution: Education and Outreach* 6, no. 18 (2013), doi.org/10.1186/1936-6434-6-18. This study reveals that the metaphors of "higher" and "lower" life forms, or "primitive" and "advanced" organisms, still surface in scientific literature. An analysis of over 60,000 biological articles found that such language persists, even in evolutionary studies that would seem to oppose these outdated ideas. These scalar tropes, reinforced by diagrams and visual representations, perpetuate a view of evolution as progressive and directional, often culminating with humans at the apex. For a critique of this notion as an oversimplification, see H. Yu "The tree of life in popular science: Assumptions, accuracy, and accessibility," in *Scientific Communication: Practices, Theories, and Pedagogies* 8, H. Yu and K. M. Northcut, eds. (Routledge, 2018), 87–107. Likewise, the Great Chain of Being continues to circulate in religious rhetorics as justification for anthropocentrism. For example, Gregory Poore, "Reconciling the Food Chain with the Great Chain of Being: A Philosopher's Reflection on Raising Sheep for Meat," in *Ecoflourishing and Virtue: Christian Perspectives Across the Disciplines*, Steven Bouma-Prediger and Nathan Carso, eds. (Routledge, 2023).
18. It is worth noting that even in recent studies of trophic levels and food webs, humans rank at a 2.21 on a scale of 1 to 5, just below the middle, sharing space with anchovies and pigs. See Sylvain Bonhommeau et al., "Eating Up the World's Food Web and the Human Trophic Level," *PNAS* 110, no. 51 (2013): 20617–20,620, doi:https://doi.org/10.1073/pnas.1305827110
19. "The Complete Work of Charles Darwin Online," *Notebook B.*, Darwin Online, 1837–1838, 74, http://darwin-online.org.uk/

20. Charles R. Darwin, *On the Origin of Species by Means of Natural Selection, or Preservation of Favoured Races in the Struggle for Life* (John Murray, 1859), 129.
21. Justine Wells, "The Energy of Whiteness," (Paper Presented at the 20th Biennial Conference of the Rhetoric Society of America, Baltimore, MD, May 2022), https://rhetoricsociety.confex.com/rhetoricsociety/2022/meetingapp.cgi/Session/1616
22. G. Evelyn Hutchinson, "Circular Causal Systems in Ecology," *Annals of the New York Academy of Sciences* 50 (1948): 221–246, doi:10.1111/j.1749-6632.1948.tb39854.x
23. Raymond Lindeman, "The Trophic-Dynamic Aspect of Ecology," *Ecology* 23, no. 4 (1942): 399–417, doi:/10.2307/1930126
24. Charles Elton, *Animal Ecology* (Macmillan, 1927).
25. Clarence Dineen, "An ecological study of a Minnesota pond," *American Midland Naturalist* 50 (1953): 349–376, doi:10.2307/2422094
26. H. T. Odum, "Trophic Structure and Productivity of Silver Springs, Florida," *Ecological Monographs* 27 (1957): 107, doi:10.2307/1948571
27. H. T. Odum, "The Biogeochemistry of Strontium: With Discussion on the Ecological Integration of Elements" (PhD diss., Yale University, 1950).
28. James E. Lovelock and Lynn Margulis, "Atmospheric homeostasis by and for the biosphere- The Gaia hypothesis," *Tellus* 26, no. 1 (1974): 2–10, doi:10.1111/j.2153-3490.1974.tb01946.x
29. Karin E. Limburg, "The Biogeochemistry of Strontium: A Review of H.T. Odum's Contributions," *Ecological Modelling* 178, no. 1–2 (2004): 31–33, doi:10.1016/j.ecolmodel.2003.12.022
30. G. E. Hutchinson, "Biogeochemistry of aluminum and certain related elements," *Quarterly Review of Biology* 18 (1943): 1–29.
31. Limburg, "The Biogeochemistry," 32.
32. H. T. Odum, "The Stability of the World Strontium Cycle," *Science* 114 (1951): 411, doi:10.1126/science.114.2964.407
33. Andrew Pilsch, "Invoking Darkness: *Skotison*, Scalar Derangement, and Inhuman Rhetoric," *Philosophy & Rhetoric* 50, no. 3 (2017): 349, doi:10.5325/philrhet.50.3.0336
34. Thorvald Sørensen, "Some ecosystemtical characteristics determined by Raunkiær's circling method In: To designate the fundamental unit of ecological plant sociology I propose the term ecotope, viz. the field delimited as an object of investigation within a given ecosystem (Tansley)." *Nordiska* (19. Skandinaviska) (Naturforskarmöteti, Helsingfors): 474–475; Arthur Tansley. *The British Isles and their vegetation.* (Cambridge University Press, 1939).
35. H. T. Odum, *Environment, Power and Society* (John Wiley & Sons, 1971).
36. E. Odum, "The Mesocosm."

37. Larsen and Johnson, *Being Together*, 4.
38. Ibid.
39. Ibid., 17.
40. Ibid., 3.
41. Ibid., 3. See also Vine Deloria, Jr. and Daniel Wildcat, *Power and Place: Indian Education in America* (Fulcrum Resources, 2001).
42. Timothy Clark, *Ecocriticism on the Edge: The Anthropocene as a Threshold Concept* (Bloomsbury, 2015); Zach Horton, "The Trans-Scalar Challenge of Ecology," *Interdisciplinary Studies in Literature and Environment* 26, no. 1 (2019): 5–26, doi:10.1093/isle/isy079; Zach Horton, *The Cosmic Zoom: Scale, Knowledge, and Mediation* (University of Chicago Press, 2021); Timothy Morton, *Dark Ecology: For a Logic of Future Coexistence* (Columbia University Press, 2016); Timothy Morton, *Hyperobjects: Philosophy and Ecology after the End of the World* (University of Minnesota Press, 2013); Bruno Latour, "Anti-Zoom," in *Scale in Literature and Culture*, Michael Tavel Clarke and David Wittenberg, eds. (Palgrave Macmillan, 2017), 93–101; Derek Woods, "Scale in Ecological Science Writing," in *The Routledge Handbook of Ecocriticism and Environmental Communication*, Scott Slovic et al., eds. (Routledge, 2019) 118–128; Derek Woods, "Scale Critique for the Anthropocene, Part Two," *New Formations* 107/108 (2022): 155–170, doi:10.3898/newf:107-8.09.2022; Joanna Zylinska *Minimal Ethics for the Anthropocene* (Open Humanities Press, 2014).
43. Joshua DiCaglio, *Scale Theory: A Nondisciplinary Inquiry* (University of Minnesota Press, 2021); Madison Jones, "(Re)placing the Rhetoric of Scale: Ecoliteracy, Networked Writing, and Memorial Mapping," in *Mediating Nature: The Role of Technology in Ecological Literacy*, ed. Sidney I. Dobrin and Sean Morey (Routledge, 2019), 79–95; Derek Mueller, *Network Sense: Methods for Visualizing a Discipline* (WAC Clearinghouse; University Press of Colorado, 2018); Pilsch, "Invoking Darkness."
44. While it is well beyond the scope of this chapter to fully engage with the longstanding conversations about scale that have taken place in geography and ecology since the 1970s, Chris Tong offers a brief history of these disciplinary engagements with scale in his essay "Ecology without Scale," examining this history in terms of OOO and ANT, and offering an approach to relationality unbounded by scale. See C. Tong, "Ecology without Scale: Unthinking the World Zoom," *Animation* 9, no. 2 (2014): 196–211, https://doi.org/10.1177/1746847714527199
45. Jenny Rice, "The rhetorical aesthetics of more: On archival magnitude," *Philosophy & Rhetoric* 50, no. 1 (2017): 27, doi:10.5325/philrhet.50.1.0026
46. Horton, "The Trans-Scalar," 13.

47. Ibid.
48. Phillips, *The Truth*, 52.
49. S. S. Qian et al., "The Implications of Simpson's Paradox for Cross-Scale Inference among Lakes," *Water Research* 163 (2019), doi:10.1016/j.watres.2019.114855
50. Odum, "Trophic Structure."
51. H. T. Odum, "Macroscopic Minimodels of Man and Nature," in *Systems Analysis and Simulation in Ecology*, ed. B. Patten (Academic Press, 1976), 250–80.
52. H. T. Odum, *Environment, Power and Society* (John Wiley & Sons, 1971), 10.
53. Laura J. Martin, "Proving Grounds: Ecological Fieldwork in the Pacific and the Materialization of Ecosystems," *Environmental History* 23, no. 3 (2018): 575, doi:10.1093/envhis/emy007
54. See Stephen Carpenter, "Microcosm Experiments Have Limited Relevance for Community and Ecosystem Ecology," *Ecology* 77, no. 3 (1996): 677–680, doi:10.2307/2265490
55. G. Evelyn Hutchinson and R. L. Lindeman, "Biological Efficiency in Succession (Abstract)." *The Bulletin of the Ecological Society of America* 22 (1941): 44, https://www.jstor.org/stable/20165143
56. Carpenter, "Microcosm Experiments," 578.
57. Ibid., 579.
58. For example, Richard Onians, *The Origins of European Thought* (Arno, 1973); William Race, "The Word *Kairos* in Greek Drama," *Transactions of the American Philological Association* 111, no.1 (1981): 97–213; Thomas Rickert, *Ambient Rhetoric: The Attunements of Rhetorical Being* (University of Pittsburgh Press, 2013).
59. H. T. Odum, "Scales of Ecological Engineering," *Ecological Engineering* 6, no. 1–3 (1996): 11–12, doi:10.1016/0925-8574(95)00049-6
60. Casey Boyle, "A Wealth of Realities," in *Rhetorical Ecologies*, ed. Sid Dobrin and Madison Jones (National Council of Teachers of English Press, 2024), 310.
61. Ibid., 311.
62. Blue Origins. "LEO Destinations." Accessed on February 2, 2025, https://www.blueorigin.com/destinations: para. 3.
63. J. G. Ballard, *The Complete Stories of J. G. Ballard* (W. W. Norton & Company, 2010).

Open Access This chapter is licensed under the terms of the Creative Commons Attribution-NonCommercial-NoDerivatives 4.0 International License (http://creativecommons.org/licenses/by-nc-nd/4.0/), which permits any noncommercial use, sharing, distribution and reproduction in any medium or format, as long as you give appropriate credit to the original author(s) and the source, provide a link to the Creative Commons license and indicate if you modified the licensed material. You do not have permission under this license to share adapted material derived from this chapter or parts of it.

The images or other third party material in this chapter are included in the chapter's Creative Commons license, unless indicated otherwise in a credit line to the material. If material is not included in the chapter's Creative Commons license and your intended use is not permitted by statutory regulation or exceeds the permitted use, you will need to obtain permission directly from the copyright holder.

CHAPTER 5

Ecology Out of Time: *Kairos* and Temporal Problems

Abstract Turning to the problem of time for environmental inquiry, this chapter investigates the ways that temporality shapes climate change and environmental crisis communication. An early focus on stability in ecological models failed to account for rapid environmental change. This chapter applies concepts like *kairos* and deep time to the evolution of temporality in ecosystems ecology, and it considers how shifting rhetorical frameworks of time can help ecological inquiry move beyond the rhetoric of crisis.

Keywords Rhetoric of time • *Kairos* • Deep time • Climate change • Ecological crisis

> We believe, fatalistically, that the ecosphere is on a straight path into catastrophe, when in some ways even the apocalypse is cyclical. […] We speak of sustainable living even as we face the specter of the Anthropocene. Our cultural vision of time is a heap of broken clocks. —Paul Huebener, *Nature's Broken Clocks*[1]

Paynes Prairie State Park, Alachua County, Florida, United States

© The Author(s) 2026
M. P. Jones, *Inventing Ecosystems*, Palgrave Studies in Media and Environmental Communication,
https://doi.org/10.1007/978-3-031-98793-9_5

Our group ambles along the paved pathway from the parking lot to the fenced entrance to the La Chua Trailhead. Volunteers from Trace Innovation, our research group at the University of Florida, arrived early to help us set up and have volunteered to help lead community participants on a guided demonstration of EcoTour, *an augmented reality walking tour of Paynes Prairie State Park. Our team of graduate students at the University of Florida had been developing* EcoTour *for months along with student collaborators across several of the digital writing courses we teach.*[2] *As our group enters the park, we see the edges of the vast savannah where wild horses and bison roam, the grasslands spreading out for miles behind an enormous live oak tree draped in thick Spanish moss. Here, I sometimes see limpkin and even owls resting in its deep shade. This event was the culmination of a multi-day symposium we organized around the topic of environmental media, featuring talks by Nichole Starosielski and Casey Boyle.* EcoTour *was a digital excursion that connects the ecological history of Paynes Prairie to current environmental issues.*

Paynes Prairie has long been a place of inspiration for environmental thinking. William Bartram wrote extensively of the place in Travels, *as well as his encounters with the Seminole tribe that inhabited these lands. Bartram's writings would go on to influence many early American environmental writers and scientists.*[3] *Just a few miles from here, H.T. Odum experimented with recycling municipal wastewater in a cypress dome. If this place could talk, it would have so many remarkable stories. Using a mobile device, visitors can view an interactive map and can scan existing signs within the park to access multimedia augmented reality overlays, including archived audio-visual media related to specific physical locations. Our goal is to engage with the park as a dynamic site by layering together the rich geological, ecological, historical, and social dimensions of the place to tell a more complete story. This is a practice known as "deep mapping," a place-based storytelling method that layers different types of information together. These methods can help convey a sense of deep time, the vast temporal expanse that typically eludes the imagination of anyone but scientists.*

In the midst of our project, Hurricane Irma made landfall in September 2017, temporarily inundating the karst topography of the area. The volume of water overloaded the Alachua Sink, the main source of drainage for the basin, flooding highways and rivers, damaging a levee, and reverting the droughted savannah into a massive lake, historically known as Alachua Lake (from the Timucuan word "Chua," meaning roughly "sinkhole"). Fish and alligators swam over what had been hiking trails. As a result, our team had to adjust the path of our proposed walking tour to accommodate the changed landscape. Despite these logistical challenges, these environmental changes offered our team a unique opportunity to visualize some of the geological, ecological, and historical elements of Paynes Prairie in real time. As our group nears the boardwalk, we spot another example of Irma's wake: a snail kite flying far off in the cloudless sky. The hurricane brought major changes not only to the physical landscape but also to the species that call that landscape home. The flooded basin has drawn the giant apple snail, a recently introduced species in Florida that is native to South America, and with it has come the endangered snail kite, whose numbers have been steadily increasing since the giant apple snail arrived a few years ago. As we continue our detoured path, I am left thinking about how the layers of place that we perceive are but one part of deep temporal rhythms, moments of quiet stability punctuated by episodes of extreme change.

The Problem of Time

Ecological inquiry is vexed by rhetorical problems of time. From early ecological models of succession to contemporary debates about the Anthropocene, environmental thought has often turned to the rhythms, durations, and thresholds of environments through a variety of temporalities. This chapter investigates the temporal dimensions of ecological research, focusing on how the Odums and their contemporaries dealt with issues of time and temporality in framing ecosystems. Namely, this chapter examines the role of time in debates around the relationship between equilibrium, stability, and complexity in ecosystems, building directly from the influence of Clementian organicism on the prominent ecologist G. Evelyn Hutchinson, who served as H. T. Odum's mentor and whose thinking influenced many ecosystems ecologists. In inventing and refining the concept of ecosystems, these ecologists sought to explain how environments could become stable communities over time. Likewise, this chapter investigates the relationships between ecology and technology through rhetorical concepts like *chronos* and *kairos*, as well as other temporal frameworks such as "machine time,"[4] "deep time,"[5] and the proposed (now rejected) geological period of time known as the Anthropocene and its many variations within EH.[6] In doing so, this chapter examines not only the relationship between ecosystems ecology's reliance on radioecology in the search for stability but also the rhetoric of crisis, introduced in part by the "Atomic Age" that today pervades environmental inquiry.

By examining the temporal frames used by ecosystems ecologists, this chapter addresses the rhetorical implications of time in ecological thinking, arguing that time is a central concern for rhetorical ecologies. Recent work in RWCS has identified what scholars are calling a "temporal turn," describing a renewed attention to the ways that time shapes rhetorical theories, methods, and practices.[7] Time is more than a backdrop for communication, as Samantha Senda-Cook and her coauthors articulate. They show how temporalities shape and constrain environmental communication, such as in crisis communication or sharing information about climate change. Their work underscores how multiple temporalities—from the symbolic to the material to the ontological—converge in environmental discourse, compelling scholars to interrogate the dominant linear frameworks that often inform action and policy. In doing so, they join a growing body of work that examines time as a rhetorical construct, offering path ways to practicing environmental communication as a place to resist and rethink dominant linear frameworks. This chapter builds on that insight, focusing on how time functions rhetorically in ecology, from its early methodological foundations to contemporary challenges in climate science and environmental justice.

Ecology as a field of study and a discipline has always relied on negotiations between a variety of temporal scales, from deep time to immediate crisis, and this creates various rhetorical challenges. From the instantaneous obliteration of a mountainside in a mudslide to the geologic accretion of sand on a beach, ecological processes unfold across a vast array of temporal scales. At the same time, ecology has struggled to account for temporalities. Arthur Tansley forwarded the ecosystem concept as a means to critique the simplistic and deterministic models ecology had relied on in organicist concepts like succession, stability, and equilibrium. At the same time, H. T. Odum's work with energy, feedback loops, and the "maximum power principle" imported many of these ecological concepts back into ecosystems ecology.[8] These problems of time affect both ecology and rhetorical ecologies, and this creates an exigency in both RWCS scholarship and ecological science, to recognize and ameliorate these temporal rhetorical problems. In turn, these solutions will better equip environmental communicators with tools to address contemporary issues like climate change, where massive timescales of carbon cycles and species adaptation collide with immediate urgencies of extreme weather, rising sea levels, changes in atmospheric and oceanic composition, and mass extinction.

At the same time, scholars in EH have increasingly turned toward the ways that the frameworks of time are not neutral entities, where people and places are affected by the different ways that timescapes are constructed. For instance, Rob Nixon demonstrates how what he terms "slow violence" obscures the shifting baselines caused by anthropogenic climate change, pollution, and habitat destruction in favor of the spectacles of disasters and crises that align with short-term political and media cycles.[9] In contrast, slow violence is "a violence that is neither spectacular nor instantaneous, but rather incremental and accretive, its calamitous repercussions playing out across a range of temporal scales."[10] For example, Jeffrey Sanders demonstrates the temporalities through which radiation moved from bomb testing sites to the "bones, blood, and thyroids of children living in the United States."[11] He details how each year from 1951 through 1958: "[A]s a parade of successive pink fallout clouds moved east in widening arcs over the United States, downwinders across the continent had been mostly left to their own devices, reckoning with the fallout."[12] Yet, Eugene P. Odum writes in his 1957 article, "Ecology and the Atomic Age," that "the atomic age can well provide the means of solving the very problems it creates."[13] Atomic energy and radiation presented new problems of temporalities that ecosystems ecology was poised to solve. Odum framed the threats posed by radiation as a "prospect [that] is rapidly transforming ecology from a rather obscure and ill-defined member of the biological family into a more organized and coherent division."[14]

Likewise, as Mark Rifkin demonstrates in *Beyond Settler Time*, the temporalities of settler colonialism present additional dangers and problems to Native peoples. In what he refers to as a "double bind," Native peoples are either "consigned to the past, or they are inserted into a present defined on non-native terms."[15] As Rifkin explains, colonialism, capitalism, and technocracy impose specific temporalities on ecological knowledge and management. For example, the temporal frameworks that underpin ecological science and environmental policy—which tend to view time as linear and progressive—are often at odds with other temporalities, such as cyclical, relational, and place-based. In response, Rifkin builds from the post-Einsteinian notions of time and relativity within physics in order to illustrate how "Indigenous forms of time push against the imperatives of settler sovereignty."[16] Along similar lines, RWCS scholars have examined the role of temporal rhetoric in Native American decolonial advocacy, examining the ways that environmentalism confronts and wrestles with similar "chrono-logics" at work in these temporal regimes,[17] as well as documenting the resistance to nuclear waste siting.[18] This chapter places these critiques in conversation with the problems of time in ecosystems ecology to demonstrate how rhetorical and ecological inquiry can work together to imagine more just, equitable, and ecologically responsive temporalities.

KAIROS AND THE ECOSYSTEM

When do ecosystems begin and when do they end? How do ecosystems form? How are they sustained in time? These are some of the questions of time that early ecologists sought to answer. As ecosystems ecology developed through the influence of cybernetics and systems theory, early models suggested that ecosystems functioned though a series of feedback loops that moved toward complexity and stability. Technomorphism brought a mechanistic view of time as a controllable variable, a temporal slice. It is useful here to turn to the interdisciplinary histories that ecosystems ecology and rhetorical ecologies share with a particular framework of time: *kairos*—or the opportune moment—and I examine its relevance to ecological and rhetorical studies. While a recent emphasis on *kairos* in RWCS has opened productive and important directions for rhetorical scholarship, such a turn can also serve to lock rhetoric into a presentist timescale, where rhetoricians are always engaging with the ever-emergent moment. The rhetorical concept of *kairos* is often defined as "opportunity" (i.e., "the right place at the right time"), the moment a rhetor may seize to persuade an audience. In Greek mythology, Καιρός personified luck and was often portrayed as young and beautiful, standing on wing-footed tiptoes as he perpetually runs forward, carrying either a razor or scales—suggesting balance, equilibrium, due measure, and the narrow margins of opportunity. *Kairos* is associated with seasonal time, as well as qualitative temporality. Often, *kairos* is defined by RWCS scholars along the lines of "the opportune moment." Yet, a survey of the last few decades of RWCS scholarship reveals the concept is much more elusive, complex, and convoluted than "opportunity" suggests.

For instance, the difference between "opportunity" and "opportunism"[19] might rely on how an audience perceives a rhetor and their discourse at any given moment, which is well beyond the capacity of an individual to control, while recent feminist readings of the concept's history suggest a definition of *kairos* as "balance and evenings," which emphasizes it as "grounded in embodiment, materiality, balance, and due measure."[20] Numerous recent scholars have argued that the power of *kairos* lies in its ambiguity.[21] The term disrupts traditional Bitzerian notions of the rhetorical "situation" as something static that can be mastered. Instead, a focus on *kairos* has moved scholars toward studying the complexity of ecologies,[22] environments,[23] circulation,[24] and the queer affect of networked bodies.[25] These theories situate rhetoric's movements within the complex flux of systems, from the scales of bodies to publics to environments to networks. *Kairos* challenges linear and chronological temporal frameworks by emphasizing the contingency of timing, as well as the ever-shifting and unstable grounds of rhetoric. Today, we can observe *kairos* across environmental discourse, such as in discussions of "tipping points" for both climate change and for biodiversity loss and conservation efforts. *Kairos* is a central component of contemporary environmental thinking because it has long played an important role in ecological science. In much of our political and environmental discourse, the temporal rhetoric of *kairos* is used as an exigency, such as when climate advocates urge that we have only ten years left to act on climate change, a message that has been repeated for many decades. These strategies have often fallen flat, as environmental rhetoricians attempt to manufacture urgency through a kairotic ecological moment. In many ways, these failures are part of the problems of time that afflict all of environmental inquiry. Just as spatial scale distorts connections to place, so do different temporal frameworks—such as evolutionary time, geological time, and proposed epochs like the Anthropocene—shape how we perceive ecological change, often obscuring anthropogenic environmental problems and reinforcing narratives of crisis, progress, or determinism.

Ecosystems ecology's emphasis on trophic mapping fundamentally requires a displacement of time that is evident in the early criticism of ecosystems ecology by evolutionary ecology.[26] Evolutionary ecologists argued that ecosystems ecology relied on a presentist model that was unable to explain the diachronic (developing over time) factors of evolution. Accounting for evolution was simply beyond the scope of a discipline already mired in dealing with vast complexity. The discipline's inability to account for diachronic time—the accumulation of rhetorical agency in evolution—proved too taxing a complexity for researchers to justify in a time of tight university budgets.[27] Studying ecosystems through synchronic (existing in a given moment) time reduced the complexity of evolution through spatiotemporal slice. Understanding the emergence of this method from such a sociohistorical moment suggests ways that *kairos* can help disciplinary histories to account for the emergence of rhetorical fundamentals that shape contemporary practices. Put differently, the concept of *kairos* allows us to see how spatiotemporal concerns are not only part of the theory but also the material history of ecosystems ecology. This history brings the connections between ecosystems and the violence of militarized colonialism into clearer focus. By understanding how temporal scales can limit ecological inquiry, we can better address these problems in terms of both their historical and future consequences.

Patrick Kangas explains that "the development of atomic bombs" would ultimately "fundamentally [change] the state of the art of Ecology," bringing it from a "small, new field" to the vibrant discipline that it is today.[28] The bomb presented a kairotic moment in the development of ecosystems ecology because it created a massive funding opportunity from the AEC while also enabling large-scale quantitative research projects into ecosystem dynamics. Radioecology aided ecologists in the search for answers to questions of self-organization, scale, stability, and complexity as well as older Clementsian concepts like succession, climax, and resilience. As major disturbances were produced in places like the Marshall Islands by the detonation of nuclear bombs and in Puerto Rico through controlled radioecology experiments, the dynamic ways that systems self-organize were brought into view, as if artificially speeding up deep or geological time processes. At the same time, nuclear radiation presented perspectives of deep time to ecological thinking by linking the immediate interventions made by military and ecological testing with the persistence of radiation in environments over massive spans of time.

While the recent emphasis on *kairos* has opened productive and important directions for RWCS scholarship, it can also similarly serve to lock rhetoric into a presentist timescale, where rhetoricians are always engaging with the emergent moment. John R. Gallagher identifies the problem of separating *kairos* from *chronos* because "neither [term] fully describes the situation."[29] His resolution is "machine time," which he defines as "mutually constitutive elements of a unified model of rhetorical time that emphasizes quantitative and qualitative perspectives together."[30] The machine time of ecosystems can likewise serve to lock ecological inquiry into a particular type of temporality. As Miller demonstrates, *kairos* is "central to the rhetoric of the scientific article" because it both interrupts time (in seeking reproducible results) and opens space (by identifying gaps in scholarship) for scientific work.[31] Building from Kuhn, Miller demonstrates how the concept participates in paradigm shifts by producing a "tension [...] between novelty and tradition," which "opens up a 'problem space,' a kairotic opportunity for scientific work."[32] By intervening in an intellectual gap, scientific knowledge develops by locating social/professional space and then inventing new intellectual space. Yet, as a kairotic space/time, ecosystems naturalize machine time by altering our sense of temporal scales using technomorphic rhetoric.

From Anthropocene to Technocene Rhetorics

Our contemporary ecological paradigm emerged from rhetorical responses to the exigencies of what many now call the "Anthropocene," a proposed geological term describing humanity's large-scale impact on the planet's fossil record. Anthropocene is derived from the Greek words *anthropos*, meaning "human," and *kainos*, meaning "novel" or "new." While posthumanist scholars (such as Donna Haraway) have frequently criticized the first part of the concept for centering the human, others have examined the issues that the second term presents in centering measurable progressive improvement that reveals what John McGuire refers to as a "myopia of innovation."[33] In the age of the Anthropocene, human activity and the nonhuman world are thought to become deeply entangled, as do the past, present, and future. At the same time, the novelty of innovation centers upon the sort of technocratic optimism associated with ecosystems ecology. Although the Anthropocene has no agreed-on start date (nor do other geological epochs) and remains a contested scientific term, it is proposed that "The Epoch of Man" refers to the period in which humans became a "force of nature," altering the climate and making marks in the geological record. In 2009, the International Union of Geological Sciences established an Anthropocene working group tasked with deciding if human impacts on the earth were grounds for a unique "stratigraphic unit." The group proposed a golden spike (or starting point) of 1952, when hydrogen bomb tests were recorded in the geologic record. After fourteen years of deliberation, the proposal was denied in a vote of twelve against and four in favor of the new epoch (with three committee members abstaining and three not voting). While the Anthropocene is not an officially recognized stratigraphic term, its uptake by numerous scholars across many disciplines demonstrates that it provides a valuable way to identify and index a large-scale period of time in which human activity has drastically altered the environment and/or earth's systems, especially as a result of nuclear technology and radioactive pollution.

Linda Tuhiwai Smith demonstrates how important the different timescales of colonized and pre-colonized time are for Indigenous critique.[34] History is an important element of her decolonial work because it allows scholars to interrupt totalizing imperial narratives. In an overlapping manner, discussions surrounding the Anthropocene epoch tend to focus on the question of *chronos* (i.e., "When did climate change begin, and when will it end?"). In geology, boundary events are large-scale climatic changes that separate epochs by leaving distinctive marks in the sedimentary record. These events are transitional moments between two boundaries— traces in the earth's record that define the geologic timescale. Sometimes these marks are definitive, but many boundary events remain disputed. While some geologists proposed July 16, 1945 (the Trinity nuclear test) as the starting point of the Anthropocene, others suggest starting dates that vary widely, from the beginning of the Industrial Revolution to as far back as the Agricultural Revolution.

While the moment of nuclear acceleration might be the point of greatest human impact on the planet's geological record, other points might offer better boundary events. For instance, geographers Heather Davis and Zoe Todd argue that the Anthropocene began during colonization,[35] altering earth systems at the planetary scale,[36] while posthumanists like Donna Haraway[37] and Anna Tsing[38] argue that the Anthropocene is itself the boundary event, a mere blip on the geologic timeline. The trouble here is that Anthropocene ecologies don't fit neatly into chronologies or totalizing narratives. As recent cultural rhetorics scholars have pointed out, colonial violence rhetorically interrupts Indigenous time, further complicating the relations between ecology and the Anthropocene.[39] Along the same lines, the idea of an ecological age has likewise been tied to the invention of nuclear technology, such as in Donald Worster's influential book *Nature's Economy*, where he says that "The Age of Ecology began on the desert outside Alamogordo, New Mexico on July 16, 1945, with a dazzling fireball of light and a swelling mushroom cloud of radioactive gases."[40] Thus, ecosystems are entangled with the rhetoric of invention just as the Anthropocene offers novel ways to reimagine past, present, and future.

KAIROTIC ECOLOGIES

Although Eugene and H. T. Odum did not coin the term "ecology," their studies of ecosystems helped bring the field from a subdivision of biology to its own discipline by borrowing metaphors from economics and cybernetics. Their research made crucial contributions to the move from ecology as a technical term to that of a paradigm shift.[41] Yet, as historians of science demonstrate, it was not just metaphors that afforded these novel innovations, but nuclear technology and direct funding from the AEC through studies of radioisotopes in laboratories,[42] as well as fieldwork conducted at nuclear production sites,[43] and at nuclear weapons test sites such as the Pacific Proving Grounds.[44] This funding stream was both cause and effect of ecosystems ecology's reliance on a synchronic timescale. Understanding the development of ecosystems ecology from such a social and historical moment suggests ways that *kairos* can help disciplinary histories to account for the emergence of rhetorical fundamentals that shape contemporary practices. For example, to return to the topic of the rhetoric surrounding "invasive" species, Donnie Sackey explains in *Trespassing Natures* that "invasion as a metaphor treats environment as a time capsule. It allows us to believe that we need to preserve environment in a form that has always existed and be resistant to change."[45] That is, while temporal slicing is an important instrument in the ecologist's toolbox, the *kairos* of such ecological research should not be mistaken as a natural state. As with other scalar storytelling, *kairos* reduces the complexity of diachronic time in productive ways, but it is important to remember that it is a rhetorical device. Retracing this counterhistory makes visible the connections between colonial Anthropocene violence and the limits of ecologies as a framework for rhetorical inquiry.

As Jenny Rice explains in her landmark definition of rhetorical ecologies, "life-as-network also means that the social field is not comprised of discrete sites but from events that are shifting and moving, grafted onto and connected with other events."[46] Although she does not explicitly refer to *kairos* in the essay, Rice's description of networks directly reflects this perspective. She goes on to suggest that "these sites (the situs) are sustained by the amalgam of processes, which can be described in ecological terms of varying intensities of encounters and interactions—much like a weather system."[47] Given this metaphor, it is important to note that *kairos* refers in both ancient and modern Greek to weather, which suggests the rich connections between the "flux" of networked ecologies and kairotic timescales. The increasing reliance on Bruno Latour's ANT by rhetoricians interested in new materialism, posthumanism, and digital networks has exacerbated ecology as a kairotic framework. *Kairos* is a central component of the concept of transformation in the contingent performance of an actor network.[48] In her introduction to an interview with Latour for *Rhetoric Society Quarterly*, Linda Walsh and her co-authors describe the formation of ANT as a quasi-rhetoric that was codified as a direct response to the kairotic time of contingent performance.[49] In response to a question from Walsh about the use of *kairos* in a passage from *On the Modern Cult of the Factish Gods*,[50] Latour claims he is influenced by the theological concept (through Deleuze and Charles Péguy):

> Physical time, isochronic time, is an important scientific and technical instrumentation, but it has nothing to do with the way we live. The time in which we are, all of us, all sorts of life forms, has a different rhythm, a different way of passing. [...] So, we don't live in physical time, we are all in a different time, and inside those times, in the plural, there is one which has been largely taken up by theology for which the name *kairos* is well adjusted, which is this time where the notion of end, the notion of definitive occurrence and rupture in the passage of the ordinary customs and habits, is highlighted.[51]

Kairotic time is deeply part of rhetorical ecologies, because as Edbauer argues, "[R]hetorical situations operate within a network of lived practical consciousness."[52] As such "place becomes decoupled from the notion of situs, or fixed (series of) locations, and linked instead to the in-between en/action of events and encounters. Place becomes a space of contacts, which are always changing and never discrete."[53] This synchronic temporal displacement emphasizes embodiment and event in productive ways for rhetorical theory, but it can also reinforce and elide the ideological inheritances of nuclear colonialism as it effaces history, abstracting the violence of ecological "invention" and "discovery" at the Pacific Proving Grounds.

As this chapter demonstrates, ecological thought is shaped by many different temporal frameworks. Both ecosystems ecologists and RWCS scholars have negotiated these complexities through concepts like *kairos* and deep time. As radioecology fueled ecosystems research, the new ecology emerged from the kairotic exigency presented by atomic testing. As such, nuclear technology has played a pivotal role in shaping the temporalities of ecological research. Nuclear radiation presented an immediate threat of disaster that spurred major funding for ecological research. Radiation also provided ecologists like H. T. and Eugene Odum with new methods to study ecosystems through temporal regimes that were previously impossible. For the Odums, radiation offered an opportunity to study environments through accelerated timescales. Their research had a lasting impact on disturbance ecology, which today helps us understand how climate change will impact environments. At the same time, ecosystems can distort a sense of time and elide technocratic perspectives on natural resource management. It can also obscure the impacts of radiation and other persistent forms of pollution on the daily lives of people across the globe who are living in the wake of nuclear testing and extractive scientific research. It can also shape and limit the current beliefs and practices of ecological researchers and environmentalists. Through rhetorical perspectives on temporality, ecological inquiry is better equipped to respond to the environmental justice problems we face.

Rhetorical perspectives on time can also open avenues for ecological inquiry to work across disciplinary boundaries and cut across dominant narratives about the environmental crisis. Understanding the emergence of ecosystems against the landscape of the Atomic Age can better equip environmental communication with tools to reconsider temporality and its role in ecological inquiry. While spectacle and crisis have served as ubiquitous frames for environmental communication, a closer look at the rhetoric of time opens environmental inquiry to perspectives informed by a multiplicity of timescales, such as those of deep mapping and Indigenous temporalities. Ecology's spatiotemporal concerns produce knowledge problems when they are naturalized in conceptions of environments and systems. Indigenous critiques of settler colonial temporalities disrupt frameworks that view systems as progressive and linear and call for cyclical approaches to environmental justice work. By understanding time as a storytelling device, in both communication and science, we can better understand how it shapes political, ethical, and social elements of ecology. By attending to these knowledge problems, RWCS scholars can foster deeper forms of engagement with ecology, work that neither ignores the legacies of militarized nuclear colonialism nor operationalizes technocratic optimism.

Notes

1. Paul Huebener, *Nature's Broken Clocks: Reimagining Time in the Face of the Environmental Crisis* (New York University Press, 2020), 3.
2. Shannon Butts and Madison Jones, "Deep Mapping for Environmental Communication Design," *Communication Design Quarterly* 9, no. 1 (2021): 4–19, doi:https://doi.org/10.1145/3437000.3437001
3. See Matthew Wynn Sivils, "William Bartram's *Travels* and the Rhetoric of Ecological Communities," *ISLE: Interdisciplinary Studies in Literature and Environment* 11, no. 1 (2004): 57–70, doi:10.1093/isle/11.1.57
4. John Gallagher, "Machine Time: Unifying Chronos and Kairos in an Era of Ubiquitous Technologies," *Rhetoric Review* 39, no. 4 (2021): 522–535, doi:10.1080/07350198.2020.1805573
5. John McPhee, *Annals of the Former World* (Farrar, Straus and Giroux, 1998).
6. For example, see Donna Haraway, *Staying with the Trouble: Making Kin in the Chthulucene* (Duke University Press, 2016).
7. Samantha Senda-Cook, Danielle Endres, Stacey K. Sowards, and Bridie McGreavy, "Engaging Complex Temporalities in Environmental Rhetoric," *Frontiers in Communication* 8, no. 8 (2023): 1176887, doi:10.3389/fcomm.2023.1176887
8. H. T. Odum, *Ecological and General Systems: An Introduction to Systems Ecology* (Colorado University Press, 1994).
9. Rob Nixon, *Slow Violence and the Environmentalism of the Poor* (Harvard University Press, 2011).
10. Ibid., 2.
11. Jeffrey Sanders, "From Bomb to Bone: Children and the Politics of the Nuclear Test Ban Treaty" in *The Nature of Hope: Grassroots Organizing, Environmental Justice, and Political Change*, Jeff Crane and Char Miller, eds. (University Press of Colorado, 2019).
12. Ibid., 158.
13. Eugene Odum, "Ecology and the Atomic Age," *ASB Bulletin* 4, no. 2 (1957): 27–29.
14. Ibid.
15. Mark Rifkin, *Beyond Settler Time: Temporal Sovereignty and Indigenous Self-Determination* (Duke University Press, 2017), vii.
16. Ibid., ix.
17. Matthew Brigham and Paul Mabrey, "'The Original Homeland Security, Fighting Terrorism Since 1492:' A Public Chrono-Controversy," in *Decolonizing Native American Rhetoric: Communicating Self-Determination*, Casey Ryan Kelly and Jason Edward Black, eds. (Peter Lang, 2018), 112.
18. Danielle Endres, *Nuclear Decolonization: Indigenous Resistance to High-Level Nuclear Waste Siting* (Ohio State University Press, 2023).

19. Carolyn Miller, "Opportunity, Opportunism, and Progress: *Kairos* in the Rhetoric of Technology," *Argumentation* 8 (1994): 90, doi:10.1007/BF00710705
20. Jordynn Jack and Emma Duvall, "Reconsidering *Kairos* through the Gendered History of Weaving," *Rhetoric Society Quarterly* 54, no. 1 (2024): 52, doi:10.1080/02773945.2023.2293957
21. See James J. Brown, Jr., "Louis C. K.'s 'Weird Ethic': Kairos and Rhetoric in the Network," *Present Tense* 3, no. 1 (2013), https://www.present-tensejournal.org/volume-3/louie-c-k-s-weird-ethic-kairos-and-rhetoric-in-the-network/; Debra Hawhee, "Kairotic Encounters," in *Perspectives on Rhetorical Invention*, Janet M. Atwill and Janice M. Lauer, eds. (University of Tennessee Press, 2002), 16–35; William C. Trapani and Chandra A. Maldonado, "*Kairos*: On the Limits to Our (Rhetorical) Situation," *Rhetoric Society Quarterly* 48, no. 3 (2018): 278–286. doi:10.1080/02773945.2018.1454211
22. Jenny Edbauer, "Unframing Models of Public Distribution: From Rhetorical Situation to Rhetorical Ecologies," *Rhetoric Society Quarterly* 35, no. 4 (2005): 5–24, doi:10.1080/02773940509391320
23. Thomas Rickert, *Ambient Rhetoric: The Attunements of Rhetorical Being* (University of Pittsburgh Press, 2013).
24. Laurie Gries and Collin Gifford Brooke, eds., *Circulation, Writing, and Rhetoric* (Utah State University Press, 2018).
25. Joe Edward Hatfield, "The Queer Kairotic: Digital Transgender Suicide Memories and Ecological Rhetorical Agency," *Rhetoric Society Quarterly* 49, no. 1 (2019): 25–48, doi:10.1080/02773945.2018.1549334
26. Benjamin Golley, *A History of the Ecosystem Concept in Ecology* (Yale University Press, 1993), 5.
27. Ibid., 6.
28. Patrick Kangas, *A History of Radioecology* (Routledge, 2022), 3.
29. Gallagher, "Machine Time," 522.
30. Ibid., 523.
31. Carolyn R. Miller, "*Kairos* in the Rhetoric of Science," in *A Rhetoric of Doing: Essays Honoring James L. Kinneavy*, Steven P. Witte et al., eds. (Southern Illinois University Press, 1992), 313.
32. Ibid., 320.
33. John McGuire, "The Problem with the Anthropocene: *Kainos*, not Anthropos," *Constellations*, 30 (2023): 128, doi:10.1111/1467-8675.12686
34. Linda Tuhiwai Smith, *Decolonizing Methodologies: Research and Indigenous Peoples* (Zed Books, 1999), 24.
35. Heather Davis and Zoe Todd, "On the Importance of a Date, or Decolonizing the Anthropocene," *ACME: An International Journal for*

Critical Geographies 16, no. 4 (2017): 761–780, doi:10.14288/acme.v16i4.1539

36. Alexander Koch et al. "Earth System Impacts of the European Arrival and Great Dying in the Americas after 1492," *Quaternary Science Reviews* 207 (2019): 13–36, doi:10.1016/j.quascirev.2018.12.004
37. Donna Haraway, *Staying with the Trouble: Making Kin in the Chthulucene* (Duke University Press, 2016), 100. She argues that "our job is to make the Anthropocene as short/thin as possible and to cultivate with each other in every way imaginable epochs to come that can replenish refuge."
38. Anna Tsing, *The Mushroom at the End of the World* (Princeton University Press, 2015).
39. See Jennifer Clary-Lemon, "Gifts, Ancestors, and Relations: Notes toward an Indigenous New Materialism," *Enculturation: A Journal of Rhetoric, Writing, and Culture* 30, no. 1 (2019); David Grant, "Like Frost on a Windowpane: On the Pluriversal Possibilities of Spacetime," *Enculturation: A Journal of Rhetoric, Writing, and Culture* 31, no. 1 (2020).
40. Donald Worster, *Nature's Economy: A History of Ecological Ideas* (Cambridge University Press, 1994), 339.
41. Benjamin Golley, *A History of the Ecosystem Concept in Ecology* (Yale University Press 1993), 188.
42. Angela Creager, *Life Atomic: A History of Radioisotopes in Science and Medicine* (University of Chicago Press, 2013).
43. Stephen Bocking, *Ecologists and Environmental Politics: A History of Contemporary Ecology* (Yale University Press, 1997).
44. Laura Martin, "Proving Grounds: Ecological Fieldwork in the Pacific and the Materialization of Ecosystems," *Environmental History* 23, no. 3 (2018): 567–592, doi:10.1093/envhis/emy007; Laura Martin, *Wild by Design: The Rise of Ecological Restoration* (Harvard University Press, 2022).
45. Donnie Johnson Sackey, *Trespassing Natures: Species Migration and the Right to Space* (Ohio State University Press, 2024), 14.
46. Edbauer, "Unframing Models," 10.
47. Ibid., 12.
48. Bruno Latour, *Reassembling the Social: An Introduction to Actor Network Theory* (Oxford University Press, 2005), 35.
49. Lynda Walsh et al., "Forum: Bruno Latour on Rhetoric," *Rhetoric Society Quarterly* 47, no. 5 (2017): 406, doi:10.1080/02773945.2017.1369822
50. Bruno Latour, *On the Modern Cult of the Factish Gods* (Duke University Press, 2010), 102.
51. Walsh et al., "Forum," 410.
52. Edbauer, "Unframing Models," 5.
53. Ibid., 10.

Open Access This chapter is licensed under the terms of the Creative Commons Attribution-NonCommercial-NoDerivatives 4.0 International License (http://creativecommons.org/licenses/by-nc-nd/4.0/), which permits any noncommercial use, sharing, distribution and reproduction in any medium or format, as long as you give appropriate credit to the original author(s) and the source, provide a link to the Creative Commons license and indicate if you modified the licensed material. You do not have permission under this license to share adapted material derived from this chapter or parts of it.

The images or other third party material in this chapter are included in the chapter's Creative Commons license, unless indicated otherwise in a credit line to the material. If material is not included in the chapter's Creative Commons license and your intended use is not permitted by statutory regulation or exceeds the permitted use, you will need to obtain permission directly from the copyright holder.

CHAPTER 6

Coda: Feedback Loops

Abstract Building on the history of ecosystems and radioecology discussed throughout this book, this chapter employs feedback loops to reflect on the future of ecological inquiry. Feedback loops can help us understand how knowledge problems of ecosystems are reflexive in ecology. These feedback loops compel us toward alternative ways of thinking about rhetoric and ecology that center environmental justice. Building from examples from the DWELL Lab, this coda emphasizes the need for interdisciplinary, land-based, and justice-oriented approaches to environmental inquiry.

Keywords Environmental justice • Public humanities • Feedback loops • Interdisciplinary inquiry

> Ecology continues to be a makeshift affair. No doubt this is precisely why it seems attractive to the kind of scientist who enjoys poking around outdoors and tinkering with things to see how they work. —Dana Phillips, *The Truth of Ecology*[1]

North Woods, University of Rhode Island (URI), Kingston, Rhode Island, United States

Early one October morning, my colleague Stephanie West-Puckett and I lug our carts and heavy bags of gear to the outer edge of the North Woods. Along the roadway, maples and oaks offer vibrant brushstrokes of yellows, reds, and browns in the soft daybreak. The North Woods is an approximately 300-acre forest owned by the university on the northern edge of our main campus. The frost on the tops of cars along Flagg Road is evidence of last night's cold snap—the first of the season. Between the chilly air and the imminent arrival of our first group of students, we hurriedly unfold the tent and arrange field notebooks, snacks, pencils, and the multilingual field guides we had worked on late into the night. We are setting up for the eighth annual National Day on Writing (NDoW) celebration at URI, which Stephanie has organized since joining the faculty in 2017. NDoW was first established in 2009 by the National Council of Teachers of English (NCTE) to celebrate composition instruction, practice, and writing as lifelong endeavors that take place across all walks of life. On our campus, NDoW events serve hundreds of students, primarily from first-year writing courses, but also including advanced undergraduates, graduate students, and students from a wide range of disciplinary backgrounds. This year, we are bringing together NDoW with the National Writing Project's annual "Write Out" event by hosting an environmental writing experience. The DWELL Lab, which I direct at URI, is partnering with NDoW to provide different types of place-based writing techniques as part of a larger emerging multimedia series called the North Woods Project, *or* NWP (https://uri.edu/dwell/projects/northwoods/). NWP *celebrates the North Woods through a variety of social, cultural, ecological, and creative perspectives and through a wide range of media and events—all with the aim of raising awareness of, and engagement with, this important part of the URI community. The goal of today's activity is to expose students to interdisciplinary writing that focuses on the different ways that place interacts with composition, from recording field notes to writing poetry. Students will practice taking jottings in their notebooks during today's activities, and then they will later use these notes and the field guides to compose a poem using more than one language. Through a guided digital walking tour experience produced by DWELL, students will listen to "How Birds Got Their Song," a traditional ecological story from the Narragansett, narrated by Lynsea Montanari, Museum Educator Associate at the Tomaquag Museum.*[2]

As the first group of students begin to gather around the welcome table, I run the script I planned once more through my head. Over the next two days, we will lead over 600 students in small groups on guided tours in the North

Woods. We will visit an outdoor classroom and take a short hike to the vernal pool, an ephemeral wetland where frogs and salamanders spawn in the spring. We will also provide the option of an accessible tour of the nearby medicinal garden. At each of these stations, students will learn about the different fauna, flora, and fungi that call the North Woods, and the wider campus community, home. The guide includes words and phrases from Narragansett regional Algonquian language, Spanish, and other land, place, and environmental terms. Through the field guide, as well as our guided discussions and activities, students are invited to consider how linguistic diversity can help us achieve a deeper, more nuanced, and expansive understanding of the North Woods. These multivalent perspectives are referred to in many Indigenous traditions as "two-eyed seeing,"[3] "Traditional Ecological Knowledge,"[4] and "Indigenous ways of knowing."[5] Through this event, we acknowledge that the University of Rhode Island occupies the unceded territory of the Narragansett Nation and the Niantic People and honor the continual presence of the Narragansett Nation on this land for many thousands of years. In welcoming students to the North Woods, we endeavor to recognize the harmful legacies of settler colonialism and Indigenous genocide in the United States, while honoring the resilience and wisdom of Indigenous communities and committing ourselves to practices that support equity, environmental literacy, and cultural respect.

Tailing Rhetorical Ecologies

What is the half-life of a concept? This book has examined the deep entanglements of ecosystems ecology with nuclear proliferation from WWII into the 1970s and beyond. This history persists in shaping ecological inquiry, despite our best urges to distance ecology from its extractive and exploitative history. The nuclear disasters at Three Mile Island Nuclear Generating Station in Pennsylvania in 1979 and the Chornobyl Nuclear Power Plant in the former Soviet Union (now Ukraine) in 1986—and following the Cuban Missile Crisis in 1962 and the Limited Nuclear Test Ban treaty in 1963—fueled antinuclear activism throughout the world, helping to curtail the development of nuclear technologies. However, the current and projected energy demands of AI and other digital technologies have recently led Microsoft to enter into a twenty-year power purchase agreement with Constellation Energy, who plans to restart Three Mile Island as early as 2027 under a new name, "Crane Clean Energy Center." Technocratic optimism reverberates like the feedback loops of nuclear reactions. In systems thinking, managing feedback as a chain of positive and negative inputs and outputs is crucial to maintaining stability, but the complexity of feedback can lead to the breakdown of systems, as it did in the meltdown at Chornobyl.[6] These events have complex and far-reaching spatial and temporal effects. The long half-lives of radioactive isotopes in Chornobyl's exclusion zone render the place uninhabitable by humans for as many as 20,000 years. Radiation from the site spread across Europe and then the globe, with elevated radioactivity levels detected as far away as North America. The long-term and large-scale effects of nuclear pollution are complex and often unclear. The estimated death toll from the Chornobyl disaster ranges from 54 by a United Nations report in 2006 to upwards of 200,000 by a report from Greenpeace that same year. The long latency of certain types of cancers and other diseases makes establishing their causes difficult. The displacement of radioactive fallout is difficult to map, and radiation spreads through food and other products in complex ways. Likewise the health effects of the traumatic experience on millions of people in Belarus, Russia, and Ukraine who were exposed to unknown levels of radiation, or the hundreds of thousands who were evacuated or remained in place. While concepts like the mesocosm allow us to imagine systems as cordoned off, contained, and separate, the byproducts of nuclear production, such as uranium tailings and other radioactive waste on Indigenous Lands in the American West, demonstrate how even supposedly tightly controlled systems generate effects across vast expanses of space and time.[7] The spatial and temporal regimes of radiation also shaped the way radioecologists invented, theorized, and studied ecosystems.

In the introduction, I discussed the origins of the ecosystem concept. In this coda, I examine how the conceptual byproducts, or tailings, of ecosystems demonstrate the need for ecological inquiry to remediate its knowledge problems. Below, I put forward the DWELL Lab as an example of land-based science and environmental communication that connects the long tail of research, teaching, and service to the land-grant mission of our state flagship university, as well as the feedback loops between the institution and continuing influence of militarized colonialism on academic research.[8] At the same time, nuclear decolonization requires coalitions that foster alliances to promote environmental justice and help protect Indigenous Lands from becoming "sacrifice zones" for nuclear waste siting.[9] This context presents an exigency for environmental justice to acknowledge the responsibilities we have to better serve communities in our research, teaching, and service. For DWELL, this means learning from and with local Indigenous history and culture and situating our work within the many groups of people and lands we call home. In collaboration with numerous groups and local organizations, and careful consultation with Indigenous leaders and culture bearers, DWELL is helping foster and repair relations with local communities through participatory projects. We collaborate with groups on the front lines of environmental harm and the legacies of the colonial project, from members of the Mashpee Wampanoag Tribe who have been exposed to harmful and persistent chemicals like PFAS through their traditional tribal fishing practices, to groups who are building resilience infrastructure through stormwater mitigation in a local park. We proceed from the conviction that we must serve the people and places that compose our home.

The word *coda* is derived from the Latin *cauda*, meaning roughly "tail of an animal," originally describing the conclusion of a musical composition. Ecologies have a long conceptual tail—a coda stretching across the globe and weaving past, present, and future. Throughout this book, I have presented a conceptual history of ecosystems through what Derek Mueller refers to as the "long tail of author citation," which helps us understand how the field's growing body of research becomes burdensome, leaving "[Kenneth] Burke's parlor [...] nowadays full and teeming, more crowded than ever before."[10] Here, I seek to build from this historical focus to understand how science communication and environmental advocacy can engage in the present and future with the long tail of rhetorical ecologies[11] through digital writing technologies.[12] As Byron Hawk puts it, "Futurity is not a problem to be solved; it is an orientation to be enacted."[13] As such, this coda branches out from the trunk of the *codex*—a term derived from *caudex*, which refers to both trees and books—in order to suggest a few, necessarily limited, provocations that propound ways field histories can inform practice. Such a *tailing* reflects the residual effects of interdisciplinary encounters between rhetoric and ecology as well as the need to maintain a close pursuit of the conceptual history outlined in previous chapters.

If coda refers to the tail of an animal, then a fitting concluding image for this *codex* is the Ouroboros (οὐροβόρος), often depicted as a snake eating its own tail. In his 1952 book, *Design for a Brain*, W. Ross Ashby draws on the image of the Ouroboros as he introduces systems theory and feedback loops.[14] That same year, Ashby attended the Macy Conferences, where the influence of technocracy, cybernetics, and systems thinking was finding purchase in the imagination of G. Evelyn Hutchinson, making its way to his student H. T. Odum and moving outward through ecosystems ecology like the inputs and outputs of a fission reaction. The Ouroboros also influenced the artist M. C. Escher, whose many depictions of the Möbius strip, such as in his print *Möbius Strip I*, were influential on the design of the recycling logo by Gary Anderson, which was created for a contest as part of the first Earth Day in 1970.[15] Finis Dunaway writes that in designing the logo, "Anderson sought to visualize infinity within a closed cycle, to evoke unending temporality within a finite space."[16] The snake eating its tail betrays problems with systems in everything from a mesocosm to a recycling program to a nuclear reactor. The Ouroboros indexes conceptual problems of ecological inquiry as we realize our environments are anything but closed or infinite systems, suggesting the perennial problems that space and time present. At the same time, a land-people approach to ecology offers pathways for renewal and regeneration through the slow process of cultivating repair and resilience across spatio-temporal scales.[17] Beyond the sociohistoriographic research outlined in this book, I want to end by emphasizing how important land-based methods and practices are for rhetorical ecologies to accomplish meaningful land-people focused work.[18]

Field Notes from the DWELL Lab

After a decade of researching and teaching courses on environmental writing, media, and rhetoric, I began to wonder what I was driving at. In the face of the interlocking global catastrophes of climate change, surveillance capitalism, forever chemicals, extremist politics, sea-level rise, mass violence, health epidemics, genocide, and a revitalization of the nuclear industry, I began to question the point of reading and writing about the environment. As Richard Miller puts it in *Writing at the End of the World*, "Why bother with reading and writing when the world is so obviously going to hell?"[19] Against these feelings of ecological grief and despair,[20] I founded the Digital Writing, Environments, Location, and Localization (DWELL) Lab at the University of Rhode Island in 2020 in direct response to the overlapping crises of social and environmental justice that were laid bare during the COVID-19 pandemic.[21] In part, the pandemic has presented a crisis of communication, a symptom of our reliance on top-down models, where information trickles down from scientists, government officials, and medical professionals to members of the public. As numerous

RWCS scholars have noted, this approach often fails to engage with the specific needs of communities as it characterizes "the public" in the abstract, fostering an environment of distrust and skepticism that is detrimental to a vibrant democracy. This model fails, in part, because it does not incorporate feedback into the system. The fractured responses to the pandemic, characterized by widespread misinformation, science skepticism, and distrust in experts, underscore the urgent need for a fundamental shift in strategies and approaches to science communication, public advocacy, and social and environmental justice that better connect with the lived experiences of local communities. In response, DWELL seeks to apply an approach to public communication that is deeply rooted in both humanistic and scientific inquiry, using feedback loops to localize intractable problems as we model co-creation, addressing the specific needs and perspectives of the groups we serve through a dialogic design process.[22] We have developed methods and practices for participatory, community-engaged work that centers accessibility[23] and directly engages with the histories and lived experiences that together compose a sense of place while providing students with high-impact, interdisciplinary experiential learning opportunities.[24] Drawing from our prior work with deep mapping and augmented reality (AR) as methods, our projects combine digital maps with multimedia AR overlays, connecting users to land-based perspectives through storytelling to promote environmental justice.[25] As participants walk through one of the DWELL Lab's AR experiences, or access content through our web maps, they view videos, audio, images, and 3D models projected into the physical environment through their smartphone. These designs communicate large-scale issues like climate justice or ubiquitous environmental hazards like PFAS pollution through methods that are embodied, entimed, and emplaced.[26]

DWELL aims to create platforms for people to participate in two-way dialogues that help bridge the gap between experts and the public, using feedback loops to address large-scale issues in ways that are meaningful at the local level. Through our ongoing location-based media projects, we demonstrate how applying theories, methods, and practices drawn directly from across the humanities and sciences can be combined with emerging technologies to integrate diverse and underrepresented perspectives into the communication process and to make meaningful contributions to the public good. In *Science v. Story*, Emma Frances Bloomfield discusses DWELL as an example of participatory approaches to science communication that places audiences in "immersive environments" that help "people focus on actions and outcomes."[27] As Bloomfield points out, DWELL works with numerous groups across the state of Rhode Island and beyond to bring "stories of marginalized communities to life" in order to "amplify often silenced voices and bring recognition to disproportionate impacts, especially around climate change."[28] This requires slowing down and rethinking many of the underlying assumptions of the contemporary research institution in order to combat the harmful and extractive practices that sometimes accompany community-engaged academic research.[29] At many times, this required treating DWELL as distinctly separate from my research, teaching, and service obligations at my institution. Simultaneously, we must also mitigate strain and potential harm for our local partners, while supporting student researchers through ad hoc grant funding at the local, regional, and federal levels. This process is often piecemeal and labor intensive. To accomplish our goals, we follow the practices set out by many other scholars in RWCS and elsewhere who have been doing this work long before us[30] in order to foster spaces for future projects both within and outside of RWCS. Our lab's practices draw inspiration from examples like Max Liboiron's Civic Laboratory for Environmental Action Research (CLEAR), which centers on "humility, accountability, and openness to others" as guiding values.[31] Following this, DWELL engages with the long tail of science communication as it stretches across space and time.

As discussed throughout this book, field histories help rhetoricians to situate relationality in terms of geological (or "deep") time, revealing important ways of understanding rhetoric as well as the fields that shape it as relations. Understanding the environmental crisis as part of colonial violence helps resist the ways that ecology's spatiotemporal concerns elide and dissociate ecology and history. By turning to community-based, place-centered approaches, we can incorporate these theoretical perspectives into practice by engaging with the material conditions and lived experiences that comprise ecological problems.[32] Such engagement enables rhetorical ecologists not only to analyze and critique, but also to actively participate in and intervene upon the ecologies of our discipline.[33] Kenneth Burke's concept of "god terms" suggests how powerfully ecology shapes contemporary rhetorical inquiry. Yet, to better challenge ecological injustices and the persistent legacies of colonial violence, we must also put rhetoric to work within our local communities, directly confronting the places and histories where injustices persist. The conceptual history outlined in this book suggests ways that RWCS and ecology can learn from one another, not only in posing critique but in developing meaningful practices to confront our biggest environmental challenges.

It is telling that Burke mentions "nature," "environment," and "history" together as examples of god terms. In setting out to engage with a counterhistory of ecology, my aim is not to replace one god term with another. As I have argued, placing spatiotemporal problems that concern ecologies within counterhistories suggests ways to become better ecologists. Just as the contemporary science of ecology has brought together ecosystems and evolutionary perspectives (to enhance a sense of time) as well as trans-scalar and community-centered approaches (to enrich a sense of place), so must RWCS seek methods to address these ecological concerns. While rhetorical field histories suggest one way to study these disciplinary ecologies from within, this framework is by no means exhaustive, and I count myself lucky that I am far from alone in seeking new ways to engage with these complex issues. To my mind, the success of solving the problems of ecology through critical and historical research alone is as unlikely as transcending time and space. However, developing a richer understanding of the histories that shape our disciplinary practices will allow us to apprehend, and hopefully to work purposely toward building, better worlds within those ecologies. The next step is to bring these efforts into greater contact with the peoples that have endured and continue to survive this extractive history. This involves moving from engaging with disciplinary counterhistories through theoretical, methodological, archival, ethnographic, and self-reflective approaches to bringing those findings into richer conversation with the lived experiences of people in the present with the aim of building coalitions that will hopefully lead to better and more just futures.

As part of the long tail of science communication, feedback loops can help us address the spatiotemporal problems of rhetorical ecologies through a focus on land-based practices. Through programs that extend ecological frameworks to multimodal, participatory composition, DWELL works to address the spatiotemporal problems that ecology presents to rhetoric. Situated within rhetorical ecologies that actively engage with community experiences and histories, DWELL fosters coalitional environmental justice. It is often a slow and piecemeal approach, subject to the divergent timescales of grant funding periods, academic calendars, and the busy schedules of our local partners across many different places, contexts, and communities. DWELL serves as a long-term conduit for collaboration, negotiating different spatiotemporal scales to create projects that take root and grow at their own pace. Ecological inquiry presents fundamental challenges and opportunities for rhetoric, and engaging in interdisciplinary inquiry situated within these field histories reveals places to blend together critique with active and engaged practices. Building better futures requires coalitions that move between and among disciplinary boundaries, bringing the rhetoric of science into greater contact with ecological science, while also engaging with the long tails, and tales, of history, culture, activism, and policy. In committing to these pathways, ecological inquiry can make meaningful contributions to better and more just futures. What lies ahead necessitates thinking deeply about how the complex history of ecology compels us toward more thoughtful engagement with the worlds we build and stories we share. Ecological inquiry requires balancing meaningful engagement with problems of space and time while staying closely attuned to the places we call home. Building a better future means learning to dwell. This is only the beginning.

NOTES

1. Dana Phillips, *The Truth of Ecology* (Oxford University Press, 2003), 120.
2. A desktop version of the story walk is available on the DWELL Lab website: https://web.uri.edu/dwell/projects/north-woods/narragansett-story-walk/. For more information about the project, see Madison Jones et al., "North Woods Project: Mobilizing Digital Field Methods and Art-Based Research for Science Communication and Environmental Advocacy," *Kairos: A Journal of Rhetoric, Technology, and Pedagogy* 30, no. 1 (2025), doi: 10.7940/M330.1.TOPOI.JONES
3. The concept of *Etuaptmumk*, or "two-eyed seeing," was introduced to the academic community in the late 1990s and early 2000s by M'ikmaq Elders Albert and Murdena Marshall with Cheryl Bartlett. See A. Marshall and C. Bartlett, "Two-Eyed Seeing for Knowledge Gardening," Michael Peters, ed., *Encyclopedia of Educational Philosophy and Theory* (Springer, 2018), https://doi.org/10.1007/978-981-287-532-7_638-1
4. See Jason Collins, Kristin Arola, and Marika Seigel, "Land-People Rhetorical Ecologies," in *Rhetorical Ecologies*, ed. Sidney I. Dobrin and Madison Jones (NCTE, 2024), 181–201.
5. See Robin Wall Kimmerer, *Braiding Sweetgrass: Indigenous Wisdom, Scientific Knowledge and the Teachings of Plants* (Milkweed, 2013) for a discussion of how Indigenous ways of knowing can function in conversation with ecological and biological science to promote ecological restoration work.

6. For more information about the Chornobyl disaster and its far-reaching effects, see Kate Brown, *Manual for Survival: A Chernobyl Guide to the Future* (W. W. Norton & Company, 2019).
7. Danielle Endres, *Nuclear Decolonization: Indigenous Resistance to High-Level Nuclear Waste Siting* (Ohio State University Press, 2023), 47.
8. Max Liboiron, *Pollution Is Colonialism* (Duke University Press, 2021).
9. Endres, *Nuclear Decolonization*, 92.
10. Derek Mueller, "Grasping Rhetoric and Composition by its Long Tail: What Graphs Can Tell Us About the Field's Changing Shape," *College Composition and Communication* 64, no. 1 (2012): 214, doi: 10.58680/ccc201220866
11. Sarah Young, Simone Driessen, and Jason Pridmore, "'We Lied to You…And We'll Do It Again' — Communicating Science Via YouTube," *Kairos: A Journal of Rhetoric, Technology, and Pedagogy* 30, no. 1 (2025). Young, Driessen, and Pridmore introduce this term to recognize "that current practices and science communication offerings can prevail long beyond an initial encounter," Young, Driessen, and Pridmore, "We Lied," para. 2.
12. Sean Morey, "Becoming T@iled," in *Writing Posthumanism, Posthuman Writing*, ed. Sidney I. Dobrin (Parlor Press, 2015), 133–145. Morey forwards "tailing" as a way to engage with the relationship between digital media and posthumanism.
13. Byron Hawk, "Counter-Traditions in Ecologies of Composition: Three Models of Futurity," in *Rhetorical Ecologies*, Sidney I. Dobrin and Madison Jones, eds. (NCTE Press, 2024), 39.
14. Ross W. Ashby, *Design for a Brain: The Origin of Adaptive Behavior* (Chapman & Hall Ltd., 1952).
15. Finis Dunaway, *Seeing Green: The Use and Abuse of American Environmental Images* (University of Chicago Press, 2015), 99.
16. Ibid., 101.
17. See Kristin Arola, "A Land-Based Digital Design Rhetoric," in *Routledge Companion to Digital Writing & Rhetoric*, Jonathan Alexander and Jacqueline Rhodes, eds. (New York: Routledge, 2018); Gabriela Raquel Rios, "Cultivating Land-Based Literacies and Rhetorics," *Literacy in Composition Studies* 3, no. 1 (2015): 60–70, doi:10.21623/1.3.1.4; Endres, *Nuclear Decolonization*; and Collins, Arola, and Seigel, "Land-People Rhetorical."
18. Caroline Gottschalk Druschke, "With Whom Do We Speak? Building Transdisciplinary Collaborations in Rhetoric of Science," *Poroi* 10, no. 1 (2014), https://pubs.lib.uiowa.edu/poroi/article/3447/galley/112353/view/

19. Richard Miller, *Writing at the End of the World* (University of Pittsburgh Press, 2005), 16.
20. See Tim Jensen, *Ecologies of Guilt in Environmental Rhetorics* (Palgrave, 2019).
21. See Farhana Sultana, "Climate change, COVID-19 and the Co-Production of Injustices: A Feminist Reading of Overlapping Crises," *Social and Cultural Geography* 22 (2021): 447–460, doi:10.1080/1464936 5.2021.1910994
22. Katerina Cizek and William Uricchio, *Collective Wisdom: Co-Creating Media for Equity and Justice* (MIT Press, 2022).
23. Leah Heilig, Madison Jones, Ally Overbay, and Taylor Roberts, "Augmenting for Accessible Environments: Layering Deep Mapping, Deep Accessibility, and Community Literacy," *Communication Design Quarterly* 12, no. 1 (2024): 33–43, doi:10.1145/3655727.3655731
24. Madison Jones et al., "North Woods."
25. Shannon Butts and Madison Jones, "Deep Mapping for Environmental Communication Design," *Communication Design Quarterly* 9, no. 1 (2021): 4–19, doi:10.1145/3437000.3437001
26. DWELL's *NWP* is an example of an emplaced project that communicates climate justice issues (https://web.uri.edu/dwell/projects/north-woods/). For an example of PFAS communication, see *PFAS Kitchen*, Dir. Madison Jones, DWELL Lab / STEEP, 2024. Augmented reality game, DOI: 10.23860/pfas
27. Emma Frances Bloomfield, *Science v. Story Narrative Strategies for Science Communicators* (University of California Press, 2024), 187.
28. Ibid., 188.
29. C. G. Druschke, "From Access to Refusal: Remaking University-Community Collaboration," *Community Literacy Journal* 17, no. 1 (2022), doi:10.25148/CLJ.17.1.010656; Linda Tuhiwai Smith, *Decolonizing Methodologies: Research and Indigenous Peoples* (Zed Books, 1999); Eve Tuck and Wayne K. Yang, "R-words: Refusing Research," in *Humanizing Research: Decolonizing Qualitative Inquiry with Youth and Communities*, Django Paris and Maisha T. Winn, eds. (SAGE, 2014).
30. For example, see Cana Uluak Itchuaqiyaq, C. G. Druschke, Lauren Cagle, and Rachel Bloom-Pojar, "To Community with Care: Enacting Positive Barriers to Access as Good Relations," *Community Literacy Journal* 17, no. 1 (2022), doi:10.25148/CLJ.17.1.010652
31. CLEAR, *CLEAR Lab Book: A Living Manual of Our Values, Guidelines, and Protocols, V.03.* (Civic Laboratory for Environmental Action Research; Memorial University of Newfoundland and Labrador, 2021): 7.

32. John Ackerman, C. G. Druschke, Bridie McGreavy, and Leah Sprain, "The Skunkwork of Ecological Engagement," *Reflections* 16, no. 1 (2016): 75–95.
33. C. G. Druschke, "The Radical Insufficiency and Wily Possibilities of RSTEM," *POROI* 12, no. 2 (2017): 1–10, doi:10.13008/2151-2957.1257

Open Access This chapter is licensed under the terms of the Creative Commons Attribution-NonCommercial-NoDerivatives 4.0 International License (http://creativecommons.org/licenses/by-nc-nd/4.0/), which permits any noncommercial use, sharing, distribution and reproduction in any medium or format, as long as you give appropriate credit to the original author(s) and the source, provide a link to the Creative Commons license and indicate if you modified the licensed material. You do not have permission under this license to share adapted material derived from this chapter or parts of it.

The images or other third party material in this chapter are included in the chapter's Creative Commons license, unless indicated otherwise in a credit line to the material. If material is not included in the chapter's Creative Commons license and your intended use is not permitted by statutory regulation or exceeds the permitted use, you will need to obtain permission directly from the copyright holder.

INDEX

A
Accessibility, 139
Actor-network theory (ANT), 55
Actual (or kinetic), 53
Actuality (or energeia), 52
Afrofuturism, 107
American environmental imaginations, 46
American Nature Writing, 46
Analogies, 3
Anthropocene, 5, 117
Anthropocentrism, 4
Anthropomorphism, 4
Apartheid, 4
Apparatus, 60
Arcadian agonisms, 48
Aristotle, 52, 94
Atomic Age, 117
Atomic Energy Commission (AEC), 75
Augmented reality (AR), 139

B
Bartram, William, 46
Biosphere, 83
Biosphere 2, 22, 100
Boundary, 92
British imperialism, 3
British Romantics, 46

C
Chernobyl Nuclear Power Plant, 135
Chronos, 117
Circulation, 54
Civic Laboratory for Environmental Action Research (CLEAR), 25, 140
Clementian organicism, 117
Clementsian, 57
Climate change, 6, 118
Climate justice, 139
Closed dimensions, 93

Closed systems, 51, 81, 96, 138
Coexistence, 100
Cold War, 81
Coleridge, S.T., 46
Colonial histories, 6
Communication, 3
Community-engaged work, 139
Complexities, 81, 93, 117
Complex systems, 98
Conceptual history, 3
Counterhistories, 20
COVID-19 pandemic, 138
Crisis communication, 118
Cybernetics, 6, 50, 83

D
Darwin, Charles, 3, 48
Darwinian evolution, 54
Deep mapping, 139
Deep time, 117
Deleuze, Gilles, 8
Deterministic ecosystems, 3
Diagram, 13, 59
Digital humanities (DH), 5, 52, 94
Digital rhetoric, 20
Digital Writing, Environments, Location, and Localization (DWELL) Lab, 25, 138
Disturbance ecology, 49

E
Earth Day, 138
Ecocomposition, 6
Ecological community, 49
Ecological engineering, 77
Ecological fieldwork, 74
Ecological inquiry, 5, 117
Ecological research, 74
Ecological study, 48
Ecological systems, 49
Ecological turn, 20

Ecology, 3, 48, 49
Economics, 83
Ecosystems, 3, 74, 117, 136
  ecology, 48
  metaphor, 52
Ecotope, 99
Electrical circuits, 51
Electrical network, 53
El Yunque National Forest, 50, 74
Emergy, 54
Emmerson, R.W., 46
Energese, 50
Energy, 6, 75, 96
  circulation, 48
  exchanges, 48
  flows, 48
  rhetoric, 50
Enewetak in the Marshall Islands, 74
Environmentalist thinking, 73
Environments/environmental, 3, 48, 93, 117
  education, 48
  hazards, 139
  history, 7
  humanities (EH), 4, 52, 94, 117
  justice, 73, 136
  thinking, 46
Evolutionary discourse, 96

F
Feedback loops, 135
Field, 92
  experiments, 48
Fieldwork, 105
Flat ontology, 56
  of OOO, 56
Florida's freshwater springs, 46
Food chain, 96
Food web, 96
Freud, Sigmund, 3, 50, 52
Freudian psychology, 3
Frontier nostalgia, 81

G
Gamma radiation, 73
Ginsberg, Allen, 61
Goethe, Johann Wolfgang von, 3
Great Chain of Being, 94
Guattari, Félix, 8

H
Haeckel, Ernst, 3, 48, 96
Historia Animalium, 94
Histories, 137, 139
Holism, 4
Howard T. Odum Center for Wetlands, 22
Humboldt, Alexander von, 48
Hurston, Zora Neale, 46
Hutchinson, G. Evelyn, 50, 117, 138
Hydrology, 49

I
Imperialist agonisms, 48
Indigenous Lands, 136
Indigenous ways of knowing, 135
Infinite systems, 138
Input-output systems, 83
Invasive metaphor, 49
Invasive species, 49
Invention, 3

K
Kairos, 78, 117, 122
Kairotic moment, 122
Kennedy, George A., 53

L
Laboratory, 92
Land-based methods, 138
Latour, Bruno, 55
Lindeman, Raymond, 50, 52
Linnaeus, Carl, 48
Lotka, Alfred J., 54
Luquillo Experimental Forest, 74

M
Machine characteristics, 5
Machine time, 117
Macrocosms, 92
Macy Conferences, 50, 137
Marshall Islan, 105
Mashpee Wampanoag Tribe, 137
Maximum power, 98
  principle, 54
McClure, Michael, 61
Mesocosm system, 22, 99, 136
Metabolism, 49
Metaphors, 3
Militarized colonialism, 75
Military and ecological testing, 122
Möbius strip, 138

N
Narragansett, 134
Natural history, 47
Nature writing, 47
Networks, 3, 54, 82
New ecology, 57
North Woods Project, 134
Nuclear colonial violence, 13, 79
Nuclear decolonization, 80
Nuclear imperialism, 77
Nuclear proliferation, 75
Nuclear technology, 15, 57
Nuclear testing sites, 75

O
Odum, Eugene, 4, 74
Odum, Howard T., 4, 50, 75
Origin of Species, 96
Ouroboros (οὐροβόρος), 137

P

Pacific Proving Grounds, 105
Paradigm shift, 74
PFAS, 137
Place-based rhetorical practices, 14
Plato, 4
Potential (or stored), 53
Potentiality (or dunamis), 52
Productivity, 49
Proving grounds, 77
Puerto Rico, 74

R

Radiation, 122, 136
Radioactive fallout, 136
Radioactive isotopes, 136
Radioactive waste, 136
Radioecologists, 136
Radioecology, 117
Rain Forest Project, 74
Recycling logo, 138
Rhetorical ecologies, 78, 118, 137
Rhetorical field methods, 14
Rhetorical fieldwork, 14
Rhetorical historiography, 7
Rhetorical new materialisms (RNM), 6
Rhetoric of energy, 52
Rhetoric of science, technology, and medicine (RSTM), 5, 6
Rhetoric of time, 127
Rhetoric plays, 6
Rhetoric, writing and communication studies (RWCS), 93, 118
Romantic views of "Nature", 47, 48

S

Savannah River Ecology Laboratory, 77
Savannah River Site, 77
Scala naturae, 94
Scalar dynamics, 96
Scale, 78, 93
 criticism, 94
Science and Technology Studies (STS), 94
Scienceman, David M., 54
Second Seminole War, 47
Settler colonialism/settler colonial, 79, 135
Silver Springs study, 49
Slow violence, 119
Smuts, Jan, 4, 47
Snyder, Gary, 61
Social and historical dimensions of science, 6
Social Darwinism, 4, 47
Space and time present, 79, 138
Spaceship Earth, 105
Spatial boundaries, 93
Spatial regime, 136
Spatial slice, 79
Spatial theory, 93
Spatiotemporal scales, 138
Spatiotemporal slices, 51, 78, 121
Succession, 117
Sun Ra, 107
System, 92
 theory, 3

T

Tansley, Arthur, 3
Technocene, 5
Technocracy movement, 8, 50
Technocratic ecological thought, 106
Technocratic managerialism, 75
Technocratic optimism, 135
Technocratic rhetoric of ecosystems, 56
Technological determinism, 6
Technology, 77
Technomorphic rhetoric, 123

Technomorphism, 5
Technosphere, 83
Temporalities, 117
Temporal regimes, 136
Temporal slice, 79
Temporal turn, 118
Thermodynamics, 54
Thoreau, H.D., 46
Three ecologies, 61
Three Mile Island Nuclear Generating Station, 135
Time, 117
Tomaquag Museum, 134
Topoi, 93
Topological hierarchy, 94
Topologies, 94
Traditional Ecological Knowledge, 135

Trophic dynamics, 3
Trophic hierarchies, 47
Trophic structure, 48
Two-eyed seeing, 135

U
US space program, 100

V
Virality, 59

W
Wordsworth, William, 46

GPSR Compliance

The European Union's (EU) General Product Safety Regulation (GPSR) is a set of rules that requires consumer products to be safe and our obligations to ensure this.

If you have any concerns about our products, you can contact us on

ProductSafety@springernature.com

In case Publisher is established outside the EU, the EU authorized representative is:

Springer Nature Customer Service Center GmbH
Europaplatz 3
69115 Heidelberg, Germany

www.ingramcontent.com/pod-product-compliance
Ingram Content Group UK Ltd.
Pitfield, Milton Keynes, MK11 3LW, UK
UKHW020113240426
470311UK00007B/61